Interpreting in the Community and Workplace

Interpreting in the Community and Workplace

A Practical Teaching Guide

Mette Rudvin
Università di Bologna, Italy

Elena Tomassini
SSML Fondazione Universitaria San Pellegrino

First published 2011 by
PALGRAVE MACMILLAN

Palgrave Macmillan in the UK is an imprint of Macmillan Publishers Limited, registered in England, company number 785998, of Houndmills, Basingstoke, Hampshire RG21 6XS.

Palgrave Macmillan in the US is a division of St Martin's Press LLC, 175 Fifth Avenue, New York, NY 10010.

Palgrave Macmillan is the global academic imprint of the above companies and has companies and representatives throughout the world.

Palgrave® and Macmillan® are registered trademarks in the United States, the United Kingdom, Europe and other countries.

ISBN 978–0–230–28514–9   hardback
ISBN 978–0–230–28515–6   paperback

This book is printed on paper suitable for recycling and made from fully managed and sustained forest sources. Logging, pulping and manufacturing processes are expected to conform to the environmental regulations of the country of origin.

A catalogue record for this book is available from the British Library.

Library of Congress Cataloging-in-Publication Data
Rudvin, Mette.
Interpreting in the community and workplace: a practical teaching guide/ Mette Rudvin, Elena Tomassini.
p. cm.
Includes index.
ISBN 978–0–230–28515–6 (pbk.)
1. Translating and interpreting.   I. Tomassini, Elena.   II. Title.
P306.R83 2011
418'.02—dc22   2011004348

10 9 8 7 6 5 4 3 2 1
20 19 18 17 16 15 14 13 12 11

Printed and bound in Great Britain by
CPI Antony Rowe, Chippenham and Eastbourne

Contents

v

List of Tables

Acknowledgments

This book is dedicated to our students in Forlì, in Bologna, in Misano-Adriatico and in Milan. They have not only given us the opportunity to share with others the thoughts and ideas contained in this book, but their motivation, enthusiasm, insights, questions and hard work have deeply enriched our lives.

We would also like to thank colleagues and friends Chris Garwood, Dominic Stewart, Peter Mead and Maria Chiara Russo, who were kind enough to read through our manuscript at various stages, for their helpful comments. Lastly, we would like to thank Jill Lake, copyeditor at Palgrave Macmillan, for her valuable help and infinite patience.

Mette Rudvin has received financial support from the Norwegian Non-fiction Literature Fund. The support – and patience – of the NFFO fund has been invaluable to this project, and the book would not have been published without it. So to them also goes our gratitude.

Introduction

This book attempts to fill a gap in the literature on training interpreters in the workplace, in both public and private institutions. Although the main focus of the book is on interpreting for public institutions, commonly known as 'community interpreting' and/or 'public service interpreting', we have chosen to include interpreting for private, commercially driven institutions and enterprises, which would usually be known as 'business interpreting' and 'liaison interpreting'. We believe this is the pedagogically innovative contribution of this book.

Pedagogically speaking, the goals, strategies, skills and competencies of these otherwise disparate sectors have more in common than they have differences. Since the goal of this book is first and foremost pedagogical, we have therefore decided to put together forms of interpreting that share the same basic face-to-face format but not the same institutional context. Because the techniques are so similar, the strategies we describe in this book could be applied to many other sectors too: interpreting in diplomacy, tourism, the media, indeed any interlingual oral translation situation in a face-to-face, or conversational, context. The authors of this book are based in Italy, where they have been working as interpreter- and translator trainers for the last two decades, and therefore the default language pair is Italian–English. All of the teaching strategies we describe here, however, are general and not language-specific, and all the language-specific examples can be adapted through English to any other language combination.

The literature

The literature on community interpreting and public service interpreting has been growing steadily along with the emergence of the discipline and establishment of the profession, and there is now a wealth of material on a large number of related issues: the history of the discipline, ethics, the role of the interpreter, cross-cultural issues, quality of performance, institutional aspects, accreditation, psychological aspects, sociological and political/ideological aspects, linguistic/discoursal aspects and more. Until recently, training issues have not really been focussed on, however (as is typical of emerging disciplines), except for issues of testing and assessment (Campbell and Hale 2003; Elder *et al.* 2006; Angelelli and Jacobson 2009). Much of the literature that deals with training issues has addressed quality evaluation (see the *Critical Link* volumes especially). There are some short focussed discussions on training, such as Helge Niska's (2005) 'Training interpreters: programmes, curricula, practices' and also Downing and Tillery's (1992) *Professional Training for Community Interpreters.* Kainz, Prunc and Schögler's (2010) *Modelling the Field of Community Interpreting: Questions of Methodology in Research and Training,* dealing with different didactic models and the development of research-based curricula, also looks very promising and we believe it will be a significant contribution to the literature.

The last few years have also seen a growing interest in interpreting training in specific sectors (especially medical and legal), illustrated by the 2010 International Medical Interpreters Association medical training conference at Harvard University. The large national interpreter organizations, such as NAATI (The National Accreditation Authority for Translators and Interpreters) in Australia, offer training courses for trainee interpreters who are planning to apply for accreditation (see http://www.naati.com.au/). The new frontier of online training techniques for third-level education is being explored by people like Hanne Skaaden (2007). Countries like Canada, Australia, the UK, the US, South Africa and the Scandinavian countries now have systematic and well-organized training systems functioning at regional or even national level for interpreting in healthcare, in the legal system and in local government. The accreditation of interpreters to ensure high quality interpreting standards is another crucial point that is now being addressed at many levels in these and other countries. Pan-national schemes for the training and accreditation of legal interpreters are currently being implemented in the EU, e.g. the Grotius-Aegis project (*Building Mutual Trust: A Framework Project for Implementing EU Common Standards in Legal Interpreting and Translation (JLS/2007/JPEN/219)*).

There are, however, few textbooks that are suitable for use in the classroom, the exception being Gentile, Ozolins and Vasilakakos's 1996 publication, *Liaison Interpreting: A Handbook*, covering interpreting for public services, business and diplomacy, and more recently, Sandra Hale's valuable 2007 volume, *Community Interpreting*. There is even less, however, by way of practical literature on interpreter training. Chapter 6 in Hale (2007) is an exception: she provides succinct but excellent training suggestions in this chapter. As elsewhere in the literature, research in sign language interpreting is in many ways far ahead of spoken language interpreting, and Cynthia Roy's (2009) volume, *Innovative Practices for Teaching Sign Language Interpreters,* is one such example. Although the essays are of course directed at sign language trainers, they provide much useful information on assessment and also, for example, on the use of portfolios in the class.

The present book attempts to fill the gap by providing a practical guide for trainers of community interpreters, public service interpreters and interpreters working with dialogue interpreting in other areas, with a range of suggestions on how and what to teach interpreter trainees. We hope this book will benefit trainers who are still fumbling in what is still in many countries an undefined field of practice and research.

The profession

A gradual awareness of the need to organize language services systematically (thus acknowledging of course the underlying importance of the need for communication) has led many institutions around the world to invest in the recruitment and training of freelance or permanent staff who are able to provide these services; it has also led to a tentative clarification of the role that these professionals should embody. In the medical sector, in particular, the need for adequate communication has been recognized because of the potential dangers of misdiagnoses and the subsequent legal liability of health professionals. Issues such as informed consent and professional liability in cases of misdiagnosis or omitted diagnosis, are very serious indeed, primarily for the patient but also for the health professional and the institution. If the law prescribes that patients must receive clear and exhaustive information to be able to give their informed consent, the interpreter's/translator's role becomes vital. Where many public institutions in the past generally relied on 'informal' and ad hoc freelance interpreters, on the principle that 'knowing the language is sufficient' – often using relatives, children, friends and bilingual staff to interpret – a slow realization of the dangers involved in using non-qualified staff has begun to emerge. What is needed, then, is a professional pool of trained interpreters to cater for hospitals, businesses, clinics, schools, police stations, prisons, job centres, immigration offices, and social service institutions.

Research in, and an increasing awareness of the importance of, sign language interpreting, as well as huge strides in the academic disciplines of translation and interpreting studies, have also given impetus to a gradual but definite process of professionalization in this area. International conferences specializing in this area were organized in the 1990s (their heritage contained in the valuable *Critical Link* volumes) and the existence of the discipline was by the end of the decade an established fact. The pioneer countries that led the way in the emergence and establishment of public service and community interpreting were Australia, Canada, Sweden and the Low Countries, followed by the US and UK and a host of others. In the wake of this development then, interpreting for public institutions is also emerging as an independent academic discipline with empirical and theoretical research on technique, terminology, legal aspects, institutional aspects and cultural aspects, through a variety of methodologies.

The practical and institutional application of this discipline is today crucial in a more and more demographically complex world where high-quality professional communication in science and technology, welfare, medicine and the judiciary has become essential. With ever-growing numbers of migrants and a growing bureaucratization of public institutions, neither clients nor service providers should today rely on the informal networks of ad hoc communication channels such as friends and relatives functioning as interpreters, but rather on a well-organized, professional and safe system of implementing communication across languages, based on clear-cut and systematic training and accreditation procedures. Organization and clarification are desperately needed in terms of interpreter roles and job descriptions and it is crucial that this is developed collaboratively between training institutions and public services. Ideally, then, training institutions – be they academic, state-based (federal or local) or NGO-run – should be able to train their students to become the type of interpreters public and private institutions need in order to provide quality care in health, legal, social, commercial and other services for people who do not speak the majority language.

The genesis of this book

The authors of this book began teaching a course together in 2001 at the then School for Modern Languages for Interpreters and Translators at the

University of Bologna. At the time, the course was still called 'trattativa' ('negotiation' in Italian), the presumed setting being interpreting in the business sector. Although we had no pedagogical experience from either business or interpreting for public services, we did have a great deal of experience in conference interpreting in medical and legal settings, and in translation (both theory and practice). This proved to be an exciting challenge, as neither of us had taught this form of interpreting before; the types of situations and the registers used in a conference setting differ greatly from those used in a typical face-to-face interpreting session in the workplace. Basing the new coursework on past experience and adjusting the materials and experience acquired to the new teaching needs was a stimulating challenge, and a continuing learning experience that we believe we can share with other trainers who are starting out in this field, possibly coming from entirely different areas of language-related disciplines and with those trainers who are looking for new ideas for the classroom. We believe that many interpreter trainers, certainly in those countries where the discipline is still largely undeveloped, find themselves in a similar situation to ours – namely that of having to 're-invent' themselves from being language teachers and interpreter/ translator trainers to this more specific application of interpreter training in a field and profession that has not yet been fully established in many countries around the world.

After several years of teaching together at an institution for translators and interpreters, we then taught separately at other translation and interpreting institutions and on 'language mediation' courses in traditional modern language faculties. Although many of the students, especially those in the ordinary modern language faculties (i.e. not translators' and interpreters' schools or faculties), were unaware of the existence of the profession before starting our course, and despite all the dire predictions regarding job opportunities, many were highly enthusiastic and showed a keen desire to try their hand at it and to study further. Unsurprisingly, most of the students from modern language faculties who were unfamiliar with interpreting found it to be far more challenging than they had imagined – their preconceptions about 'interpreting being easy if you know the language' were regularly disproved, often to their amusement and sometimes to their frustration.

This mutability is also very much the spirit of our book: we would like it to be used as a blueprint that provides the trainer with theoretical and practical tools that can be used directly in the classroom. The book is written with this hesitant readership specifically in mind, to give ideas, support and practical hints to teachers who are new to this field. Indeed, each trainer's specific needs will set the tone for, as well as the limits to, classroom activities, and ultimately the 'chemistry' in the class will be the real measure of success. As all teachers know, this chemistry is related to so many unpredictable factors, not least student dynamics, but it can be enhanced by using a student-focussed approach where students engage actively in interpreting, discussions and preparatory work.

Pedagogical aims: why include the business sector?

Although this book, as most of the literature, deals primarily with interpreting in public institutions, we chose to include interpreting in the business sector for two main reasons. The first reason, suggested above, is that the technical skills acquired by the students are very similar to those used in the other areas we deal with. An interpreter who has been trained to interpret in hospitals and police stations can quite easily use those same technical interpreting skills in other areas by learning new terminology and extra-linguistic knowledge about systems and institutions. We believe, therefore, that by including the private commercial sector we can offer future interpreters a wider scope for job opportunities, not just for public institutions and commerce but for those areas mentioned above: tourism, media, diplomacy, education. The second reason is that in many countries where the profession and infrastructure are as yet undeveloped, most interpreters are forced to make ends meet by working in a variety of sectors. The business sector, for example at trade-fairs, is an area in which newly trained interpreters can quite easily find freelance work to gain experience and more competence, although it will rarely be sufficient to provide a regular salary. Nevertheless, despite their commonalities, the private commercial sector does to some extent fall outside our main area of concern in this book, namely where macro-structural aspects affect the interpreting process: issues such as institutional discourse, institutional power asymmetry and hierarchies, immigration, etc. Having said that, cross-cultural issues are also crucial in international business communication, as the vast literature on intercultural business management demonstrates.

The similarities between community and business interpreting lie, of course, in their mode – they both adopt the same range of face-to-face interpreting techniques in a consecutive, semi-consecutive or chuchotage mode (to be discussed in chapter 1). This is the other reason which has led us to include business interpreting in this book: as a pedagogical tool for face-to-face interpreting *tout court* it is useful because the material is easy to find (textbooks and other material on business language are abundant, as are language exercises); furthermore, students are often somewhat familiar with business terminology in L2 and in their own language (which is less frequently the case with health and legal terminology), and it ties in well with field-specific language teaching courses that are often taught in modern language faculties.

Who is this book for?

As mentioned, because this discipline is still so new and – in many countries – undefined, this book tries to provide tools for those trainers who find themselves in the position of teaching a completely new subject which they know little about. In many countries (such as Spain and Italy) the figure of the interpreter often merges with that of the 'language mediator' and/or 'cultural mediator', and many of the trainers in NGOs or regional institutions have no experience in interpreter teaching. Also, many of the teachers in the more generic language mediation courses, popular in some Mediterranean countries, are language teachers who have no experience in interpreting or even translation. They are often at a loss as to where to begin to construct an interpreting or 'language mediation' course. We hope that this book will be useful not just for experienced trainers in countries with organized interpreting services, but for people who find it difficult to devise courses because the role of the interpreter in their country is so ill-defined and poorly understood, respected, organized and paid that this is perforce reflected in the interpreting service, performance and quality around the country. (The students should be made aware of the situation in order to plan – realistically – their possible future as interpreters, without false expectations and hopes.) We have therefore attempted to provide a methodological format that is user-friendly and simple enough to be used also by trainers with language teaching experience, but with limited experience in this specific field.

How much experience must trainers have in interpreting techniques to make use of this book?

We have made the exercises in the book simple enough to be adopted by trainers whose background is in language training rather than interpreting and/or translating. We have attempted to explain basic interpreting techniques in such a fashion as to be accessible to trainers who do not have a specific interpreter-training background. Likewise, we hope that those trainers who do have experience in interpreting will nevertheless find the exercises and general comments useful for classroom work, and stimulating enough to generate discussions in the classroom on issues that go beyond the technical aspects of interpreting – for example, the code of ethics and cross-cultural communication problems.

Experienced teachers of community and conference interpreting or sign language interpreting and translating will of course be able to put together their own course quite easily. There are also a number of excellent training courses in medical interpreting and in legal interpreting around the world today, many of them online, that are highly specialized. This book does not address that particular niche, however, but is aimed at a more generalized training level that can accommodate a variety of sectors, using a user-friendly and accessible 'blueprint' into which other terminology can be inserted. It provides a wide range of practical-technical training formats so that each trainer can tailor a course targeted to the specific needs of the community and/or students.

How much interpreting experience or training is expected of the students?

Our own students have ranged from experienced students of conference interpreting at interpreting schools to students from modern language faculties with very little idea of what the profession and activity involves, and we hope that the book reflects this range. It can be adapted to students with varying interpreting experience (especially the practical exercises, by adjusting the level of difficulty). We have attempted to provide a platform that is both general and easy enough to suit beginners' groups but not rule out other students. With appropriate adaptation of extra material and exercises by the trainer, the book can be used at both undergraduate and MA level.

Language-specific?

One last question remains: to what extent does this book cater to language-specific courses? It is definitely true that many of the examples – and not least the course contents in chapter 3 – are taken from our teaching experience in Italian institutions, but we hope that these can be used as models on which to build language-specific courses in other language combinations. (We also briefly discuss the methodology of language-specific versus language-general courses in chapter 4.) Although our experience is primarily, but not exclusively, from third-level institutions, we believe that the ideas presented in this book can easily be adapted to courses outside the university by adjusting the level of difficulty and/or specificity. Since our default language pair is Italian L1 and English L2, our focus has been on students as L2 English speakers. However, we have constructed the examples (especially in the practice dialogues in chapter 6) in such a way as to make them adaptable to other language pairs.

Structure of the book

The first part of this book provides an introductory discussion on various theoretical aspects of the discipline: nomenclature and categorizing the discipline, the interpreting profession, interpreting modes, the translation process, the code of ethics, migration and cross-cultural communication related to public institutions and social and professional hierarchies. Chapter 1 situates the interpreting process firmly in the wider community whilst chapter 2 is a more theoretical discussion of cross-cultural interaction and the implications of communication between the different actors in the community, especially between the local community and the newer ethnic communities. The second part of this chapter provides practical information for the classroom trainer on how to put into practice cross-cultural knowledge and awareness in the interpreting session. Chapter 3 provides background material that can be used as a blueprint in the three sectors we have chosen to focus on: health, legal and business. It gives examples of how language services in the various sectors are organized, and then gives non-language-specific and non-culture-specific information on material that can be used in class to cover each of these sectors, such as settings and registers, multi-tasking, terminology, communication, interpersonal skills and cross-cultural issues. Chapter 3 contains a final section on the various interpreting modes used in different sectors and settings.

The second part of the book is a practical contribution which provides ideas on how to structure an interpreting course and a wide range of exercises to use in the classroom. Chapter 4 is a practical chapter about teaching methods that gives the reader ideas on how to put together a course (for example the structure of a course, whether it should be language-specific or generic, assessing student levels and interpreting competencies). It also provides a summary of the main skills we believe students should acquire at the end of the course, what we have called 'the a-b-c of interpreting competence'. Chapter 5 is a highly practical chapter which addresses concrete communicative issues from a peda-gogical point of view, issues such as register, terminology, pragmatics, language varieties, first or third person pronoun use, etc. It examines the advantages and disadvantages of peer-group role play or individual practice with the teachers playing out the dialogue, and many other issues. Chapter 5 is the most practical, hands-on chapter. Here we pro-vide the reader/trainer with a variety of practical exercises and helpful tips on where and how to find teaching material.

Dialogues

Chapter 6 is a collection of fifteen dialogues in the health, legal and busi-ness settings, annotated by the authors, to be used in the classroom. In our experience as trainers we have found that one of the most time-con-suming tasks – but pedagogically most effective – has been the writing of simulated dialogues for role plays to use in class. These ready-made dia-logues enable the trainer to provide such classroom practice quickly and effectively, as they can be adapted to the trainer's own needs and language combinations, thus alleviating some of the trainer's more tedious tasks as well as providing ideas with which they can build their own dialogues. The dialogues take the form of conversations in the workplace between representatives of various Italian institutions and non-Italian-speaking service applicants and/or other interlocutors. Examples are a conversa-tion between a doctor and a patient, a police officer and a witness, a courtroom scene, a meeting between Italian and a meeting between Italian and foreign businessmen discussing products etc. The dialogues are all in English/Italian, with translations of the Italian passages, but can easily be adapted to other language combinations; the trainer can easily improvise and elaborate upon them. The authors' comments provide information on terminological, pragmatic, interpersonal, mnemonic, discourse and cross-cultural features that help trainers apply these dia-logues with maximum effect to their own needs. Practical examples are crucial in generating a discussion in the classroom on field-specific communication and terminology, interpersonal communication across cultures, interpreting techniques and strategies, not least so that the stu-dents can see how an interpreting session takes place 'in real life', or at least an approximation of a 'real life' scenario.

Note: throughout the book, simply to avoid unwieldy formulations such as 'himself/herself', we will refer to students and interpreters as 'she' and other participants variously as 'he' or 'she'.

1
Interpreting In and For the Community, Between Practice and Practitioners: A Few Theoretical Premises

According to the *Universal Declaration of Human Rights*, all individuals should have equal right to access legal, social and health services; this is, metaphorically, the 'Hippocratic oath' on which our profession rests. We would also argue that the right to trade and to pursue international contacts should also be a basic civil right. With steadily increasing migration around the world, the need to communicate between groups of people and communities who speak different languages is essential not just for day-to-day cohabitation and establishing working interpersonal relationships between people from different ethnic groups living in the same territory, but also for very practical purposes such as trade, co-governance, education and the smooth running of institutional life. The sad truth, however, is that language services in much of the world are so poorly organized and undervalued that this premise is frequently flouted because one of the parties does not speak the majority language. Indeed, the lack of a common language becomes a serious obstacle to equal access to such basic services. Establishing an interpreting profession – whether we call it 'community interpreting', 'public service interpreting', 'liaison interpreting' or the broader 'language mediation' found in many Mediterranean countries – is one way of securing this basic civil right.

These terms all denote the range of professional activities we are referring to in this book and that we believe are crucial in safeguarding international and intercultural communication. This activity, or this range of activities, has come into its own as an autonomous academic discipline only in the last couple of decades. Because of the specific needs of the various institutions involved and their growing awareness of the importance of language services, it seems to us that there is a growing tendency to specify 'medical interpreting' and 'legal interpreting' rather than the more general denominator 'community or public service interpreting' and thus to distinguish clearly between areas of application. However, the varied nomenclature represents a peculiar fragmentation of both academic discipline and profession (more so than in conference interpreting) which has led to a number of terminological problems which we will be addressing later in this chapter. (See Hale 2007: ch. 1 for more information on nomenclature.)

First of all, though, let us clarify exactly what this activity consists in. By 'community' or 'public service' interpreting we mean interpreting from and into two different languages between two or more people who are physically present in an institutional or workplace setting. This form of face-to-face interpreting may be used in many settings (for business delegates, tourists, foreign students, short-term employees, people seeking residence permits, asylum seekers, cultural contacts, and so on). In community interpreting, and the closely related public service interpreting, the parties involved are almost always a person who speaks the national language and represents an authoritative institution or association (hospital, police, court, school, job centre), and a person who comes from a different country and does not speak the national language. In the above situations, this interlocutor would generally be a person from the category usually referred to as 'migrants'. Because of the close tie with the migrant population, there is no doubt that this profession is destined to grow in order to match the exponential increase in migration across the Western world, hand in hand with a general increase in international contacts and globalization.

Because migration and globalization, and therefore foreign language speakers, are so closely tied up with the administration of institutional life, the issue of communication in institutional settings becomes paramount. We believe that this aspect of interlingual communication, especially how the differences in language and cultural expressions affect communication and discourse strategies, is so important that we begin this chapter with a brief section on migration before moving on to issues of language and culture in the translation process and issues of nomenclature and terminology, concluding the chapter with a brief discussion on how to generate a positive attitude towards this profession by raising awareness among service providers about the importance of, and difficulties inherent in, this work.[1]

We have chosen, then, to dedicate this first chapter in an otherwise very practical book to discussing theoretical issues that we believe are crucial to gaining a full understanding of this profession and of this activity. This first chapter is thus   meant not only as a preliminary commentary to what follows but, we hope, will stimulate the interpreter trainer to discuss interpreter-related theoretical aspects in class with the students. We have found such discussions vital to the global training process and we have found a great interest and curiosity among the students when discussing these issues. We hope our readers will find the same.

1.1 Interpreting and migration: migration in context

It is sometimes easy to forget that the phenomenon of migration has been an intrinsic part of the organization of the human species since time immemorial. Migration is part of the DNA of the human species. Throughout history, migrants have been driven by a wide range of motives, such as an innate adventure lust, a desire to travel and discover new customs and languages, a desire to better one's lot or to seek new pastures and hunting grounds, and the desire for spiritual and cultural development. Last but not least, human beings migrate to escape from enemies, war, natural disaster and poverty. Putting migration in a wider historical, demographic and political context gives the student a platform from which to analyse the unique and sometimes fraught rapport between clients and service providers in institutional settings. Underlying this rapport is the migrant's motive for leaving the home country and the host country's perception of these motives. Economic migration towards the West has increased exponentially during the last half-century but the impact that this demographic trend has had varies greatly from country to country, depending also on whether or not they have a colonial past and established relations (as have the UK and France , for example) with the 'developing world'. Those countries with much weaker colonial traditions (e.g. Italy) tend to have less pre-existing cultural or material infrastructure to accommodate a mass influx of immigrants. Other countries with no imperialist past (e.g. Australia) have handled this demographic challenge positively despite having been themselves the subjects of colonial rule, or perhaps precisely because of it. Yet others with no imperialist past but a strong democratic tradition have also handled this challenge well (for example, the Scandinavian countries).

These issues will of course vary widely from country to country, and the trainer must tailor any such introduction to local specifics. Trainers could also examine the rapport between the local and new communities and the perception of immigration in the media and its enormous impact on relations between ethnic groups in the community.[2] Trainers working in EU countries could develop the role that migration in their country plays within the European community and relate their own local specific circumstances to other European immigration laws, traditions, and pan-national initiatives aimed at standardizing language services and training. There are a number of more specific comments on migration that we have also found helpful as a starting point for classroom discussion. (The comments are of course coloured by our own views on this phenomenon and on the connection between migrants, collectively or individually, and our local community. We hope nonetheless that they may prove useful as a point of departure for classroom discussion.)

1.1.1 Relations of asymmetry between developing and developed: givers and takers?

From the perspective of Western countries, many perceive migration as a novel phenomenon and also as a one-way phenomenon from the so-called developing to the so-called developed worlds (although many predict that this directionality may begin to change in the next decade towards the new, former developing, world superpowers, especially China and India). It is also a perspective that assumes that the migrants to the West are driven exclusively by economic self-improvement. It would be unreasonable to deny that this is by and large the case, but it is also true that the prevailing public rhetoric (for rather obvious ideological, political, historical, economic and natural-geographic reasons that would be far beyond the scope of this book to discuss) adds to this simple demographic fact an element of deep asymmetry. By asymmetry we mean that migration is largely perceived as a phenomenon by which developing world 'takers' move to the developed modern world 'givers', the former being indebted to the latter through receiving housing, jobs, schooling, healthcare services, legal protection, transport and access to various other goods. In short, the industrialized world provides for, assists and protects the 'new' immigrants. We believe that this is a somewhat skewed, or at least superficial, view of the migration phenomenon as a whole. Furthermore, the resulting media discourse is constantly generating an a priori asymmetry between the host community and migrant community.

In what way, one might ask, does this affect community interpreting? The first point we would like to make, backtracking to the beginning of this section, is an obvious one: the rapid and significant increase of migration to the West in the last few decades has led to a need for communication between the speakers of the host country language and arrivals to the host country whose aggregate discourse communities consist in hundreds of languages. There are myriads of communicative interfaces between the host and new communities in both the private and public spheres, ranging from interpersonal relations between neighbours, school children, colleagues and 'shop talk' in the private sphere, to schools, hospitals, legal institutions and employment services in the public sphere. All of these instances of communicative interfacing will present a variety of language- and culture-related challenges in varying degrees of quantity and importance. There are few opportunities to co-ordinate or manage such communication in the private sphere, precisely because it is private. In the public sphere, however, communication is suddenly 'everybody's business', for a number of reasons. To ensure the smooth running of a large community, governance and institutional life must function effectively, and effective communication is essential to that functioning. There is a great deal at stake for the community as a whole in the management of public affairs, not least in the economic sphere. Local and national governance is achieved through the tax-payer's money but also through the citizen's electoral vote, so citizens naturally feel they should have a say in how affairs of the state are run.

The role and organization of interpreters in healthcare, in education, in the legal and social settings, vary greatly precisely because they reflect the host country's perception of the phenomenon of migration and of the direct and indirect experience of living with these new communities. The public perception of migration as a global or national phenomenon and the perception of the migrant's level of 'indebtedness' to the host country will affect how services implemented to ensure effective communication are organized, the budget and resources allocated to them, and so on. Unfortunately, when it comes to language services, the general perception of asymmetry that we have described above is exacerbated by a very general and profoundly simplistic view of verbal communication as a mechanical process. If language is a mechanical process, the layman may deduce, then both language learning and translation between languages are simple matters, and can easily be resolved given a basic knowledge of vocabulary/grammar and a glossary. These two perceptions have led to an enormous undervaluing of the complexities both of language as a cultural artefact and of translation (also oral translation), as a profoundly complex cognitive and cultural process. If there is no real understanding of, and no pressure from the general public to address the complex issues of, cross-cultural language communication, it is unlikely that these services will be provided or budgeted for (public institutions after all mirror the public's needs and wishes). This is what is happening in many countries in Europe and will only change, we believe, when supra-national organizations like the EU provide directives and sanctions in this area (this is beginning to happen in the legal area) and/or when professionals in public institutions themselves see the necessity for effective communication to avoid costly mistakes such as medical misdiagnoses and mis-trials in court.

Any form of institutional communication will be affected by institutional norms and hierarchies, but the particular relationship that pertains to host and migrant, for both 'political', but also culture-related and language-related reasons, in an institutional communicative event will be fraught with difficulties and challenges – also at the micro-linguistic level – that are seen as the responsibility of the translator/interpreter. Ideally, this responsibility should be shared by institution and translator/interpreter in a collaborative fashion that respects the inherent challenges of language/culture transfer and the specific needs of the institution. There are, then, two spin-offs of the above discussion that affect more directly our work as language-service organizers and interpreter trainers. The first is the need to include in a teaching curriculum supra-linguistic aspects related to host and migrant cultures and to the host country's own specific migration history. The second is awareness of how the pervasive asymmetry between host and migrant is played out, not only in the social, cultural and educational institutions, but also through the institutional asymmetry which pervades the very discourse strategies used by host and migrant. It is worth re-stating that the urgent need for migrant communities to understand and be understood by public institutions on which they rely for most basic services needs to be addressed efficiently and realistically within the limitations of the economic and organizational parameters of public institutions.

1.1.2   Eurocentric bias

Although this discussion on the connection between migration and interpreting, and in particular the focus on asymmetry between developing/developed nations and internal institutional asymmetries, reflects what is happening in the profession today and especially what is represented in the literature, it has a strong eurocentric focus. There are two reasons for this, in addition to the obvious reason that we write from where we stand as professionals and trainers, where we have gained most of our experience and where our training for students is located. The first is that the countries that have led the way in the development and establishment of this discipline and the organization of the profession are first and foremost Western countries such as Australia, Canada, the US, the UK and the Scandinavian countries. Adequate financial resources and a generally positive approach towards immigration are important variables here. The second reason – a spin-off of the first – is that much of the literature in the field stems from, and reports on, Western countries, although there are fortunately many exceptions (for example, studies on South Africa, China, Malaysia or on these ethnic communities in Western countries). Important international conferences on community interpreting, translation and multiculturalism have been organized in South Africa, China and Iran, and these are also positive signs.

We would like to redress a tiny bit the bias in our book by reminding the readers that much of the cross-cultural communication that is taking place around the world as we write takes place outside the Western world. Countries and regions such as Singapore, Hong Kong, Malaysia, Indonesia, the Emirates, Brazil, Egypt, China, India, South Africa, and many more, have a wide range and mix of ethnic communities (many have enormous ethnic labourer communities) that lead to complex language and communication issues that require effective solutions. In many of these countries, however, people speak 3 or 4 languages, so the communication needs will not necessarily be the same as in Western countries. Also, with the booming Asian economies producing and the West increasingly consuming their products, the demographic and power balances may begin to shift. Hopefully, the development of the discipline and the emerging literature will begin to give a more realistic representation of language communities around the word. If this happens, new and different issues will clearly emerge – for example, the dialectical use of global languages such as Chinese and Arabic, English and French in Asia and Africa, and the political and symbolic use of various hegemonic majority languages (potentially affecting institutional asymmetry in very different ways from those mentioned above).

1.2   Language, culture and the translation process

Given that language is a reflection of the culture it 'speaks', and given that all cultures (and sub-cultures) differ, the language (or language variety) through which culture is expressed cannot, logically, fully match any other language. This applies to virtually all areas of life (food, science, religion, art, music) but also to legal systems, governance, health systems, education systems and public administration. The Critical Discourse analyst Norman Fairclough (see especially 1995) argues that language always functions as a representation of the world: it is an experience of the world, rather than a pre-existing code. Texts, then, are systems of knowledge and belief. Text production (by writers, translators, speakers, interpreters) is not 'just language', but social interaction between participants in a verbal exchange, in discourse. Subjects belonging to the same group have shared and agreed-upon knowledge bases: they share a knowledge of language codes, knowledge of principles and norms of language use, knowledge of situation – in short, knowledge of the world. Verbal interaction must therefore be one of several modes of social action and one that presupposes a shared knowledge of structures (social, situational, language codes, language norms). When communities that do not share this same knowledge of structures meet, communication problems arise and a shared interface of knowledge must be constructed. Because it is so crucial to the process of interlingual and intercultural communication, this point is worth spending a few lines on. It is indeed essential that students understand this underlying process of the complexity and almost insuperable challenge of language transfer and that they are mature enough to make independent translation decisions when there is no one-to-one correspondence between the languages they happen to be working with.

The most skilled and professional interpreter will not be able to force this linguistic and cultural mis-match to match – indeed, it is that very difference and variety in both language and culture which enriches mankind's diversity. For a translator or an interpreter, whose task it is to find correspondences, this is perhaps the most frustrating – but stimulating – dilemma they must face. Often, an interpreter or translator will have to coin or invent correspondences where they do not exist and one of the most difficult tasks of the interpreter, then, is to find a balance between the 'idea' contained in the source text, the original, and the reformulation of that same 'idea' in a language that does not necessarily possess the instruments to express that same 'idea' – indeed, it may not even possess the 'idea' itself. (There is another significant leap of faith here – we are working on the assumption that the interpreter/translator understands the speaker's/writer's 'idea' in the same way. See chapter 2, note 4.) In any case, the translator's task resides in creating a balance between the two 'opposing forces' of the original and the translation. Since a translation can never be an exact equivalent of the original, the translator will need to resort to various strategies to communicate to the reader the gist/meaning/pragmatic effect of the original utterance.[3]

In an oral text this becomes even more difficult because there are so many paralinguistic aspects involved. What do we mean by finding that 'balance' in an oral translation? Is it a question of deciding to be 'closer to the original utterance, language and culture' or 'closer to the target language and culture' when recreating the utterance?[4] Furthermore, is the interpreter permitted to adjust the politeness register to suit the target situation? Is she permitted to explain how symptoms are expressed through a different terminological and taxonomic system that has to do with a particular vision of illness, of the body, of gender? Does translating a communicative event involve taking into account that avoiding eye-contact is respectful rather than disrespectful? Should translation strategies always be governed by the purpose of the specific situation? These issues are the very foundations of translation/interpreting studies and we can only allude to them here. In the classroom they are crucial, but abstract concepts are open to misunderstanding and confusion. It is important therefore to deal with them, especially during role-play simulations, using practical examples.

The degree to which an interpreter is permitted to actively participate in the communicative event as a proactive agent is hotly debated in the field, we believe that because of the above-mentioned intrinsic and complex bond between language and culture/society, an interpreter or a translator would not be fulfilling their mandate to translate accurately were they to ignore that bond. Having said that, we certainly do not advocate a 'free-for-all' translation where the interpreter subverts the communication goal as a result of her own private agenda, linguistic or pragmatic incompetence, or her inability to co-ordinate the speakers, due to inexperience or lack of competence or even uncritically at the service of prevailing host/target culture norms.

Hale (2007) addresses this dilemma very sensibly and clearly in chapter 2 of her book, where she discusses 'direct' versus 'mediated' approaches. Perhaps it is not always easy for the interpreter to make this distinction so clearly in her everyday working life, but it is important for interpreters that this distinction becomes an internalized process of self-reflection and self-analysis. There is now a substantial body of literature on the role of the interpreter from the point of view of interpreter participation (Wadensjö 1998 and Angelelli 2004b are particularly prominent in this debate, in particular regarding the 'visibility' of the interpreter in the communicative act), a discussion that mirrors wider trends in the social sciences and translation studies (see Rudvin 2002 and 2007 for more details). The students should be taught not to uncritically exclude either approach, but rather to be aware of the different methodologies among practitioners that derive from different disciplinary/ideological/historical standpoints, as well as from the specific needs of the context, institutions and service providers involved. It is equally important that the students are taught how dangerous it can be not to respect accuracy, not only at the terminological level but also at a more encompassing communicative level. A useful way of envisioning and explaining this dilemma which is of paramount importance to interpreting studies is as a continuum from a mechanical to a pro-active approach adopting strategies appropriate to different points on the continuum.

1.3 Classifying by technique and areas of application: terminological confusion

As mentioned above, the form(s) of interpreting that we are dealing with in this book, go(es) by a number of different labels, within and across national borders. We have seen that there are a number of reasons for this lack of terminological homogeneity, most importantly because the discipline is so new. Another more practical reason is that interpreters in most countries will very often work in a number of different sectors, making it difficult to make clear-cut distinctions between health/legal/social service/business interpreters (although, as mentioned, this trend seems to be changing in countries such as the US, Canada and the UK, with more organized interpreting services). It is worth noting that the terms we have been using so far denote the specific *sectors* and institutions in the community in which interpreting is provided. There are other ways of classifying this activity however, and one of these is by denoting the method, or mode, of interpreting. We have found that students find it useful to clarify these differences at the beginning of the course as it helps them understand better how this profession works on the ground and how the various categories, that are often used somewhat indiscriminately, overlap.

Many definitions of public service – and of community – interpreting have been proposed in the literature, and they are very often extremely generalized. For example, the Federation of Interpreters and Translators (FIT)'s Community-Based Interpreting Committee has given the following definition: 'CBI encompasses any interpreting which takes place in everyday or emergency situations in the community. Possible settings include health, education, social services, legal and business' (www.fit-cbi.org; site no longer accessible).

We find the following situation-based definition of 'community interpreting' useful (Cambridge 1999: 209, based on Garber, adapted here) because it addresses macro-linguistic factors that we believe are important, not least cultural and institutional asymmetry, and gives students a more vivid picture of interpreting as it is performed 'on the ground':

• The setting involves an interview between a service provider and someone who needs or wants the services (a 'client').

- The interview arises out of some sort of crisis in the life of the client.
- There is a significant level of risk inherent in the situation (this will apply only in some cases).
- Cultural differences between service provider and the client increase the risk, also the risk of miscommunication leading to misdiagnoses or other serious consequences.
- There is a power imbalance between provider and client.

We have already alluded to the distinction between 'community interpreting' and 'public service interpreting', the latter referring specifically to public institutions which are an intrinsic part of the community. One might argue that the two should thus be interchangeable, but many scholars and practitioners object, for example, to the categorization of court interpreting as a form of community interpreting, perhaps because of the status embedded in the institutional weight of the courts. 'Community' interpreting may also be adopted in areas that are not strictly public. Furthermore, the term also has strong connotations of social services and 'assistance' which may affect both the institution's and the interpreter's perception of the role of the interpreter and/or client.

1.3.1   Dialogue and face-to-face interpreting

We have mentioned two other terms that fall outside the categories that denote setting, namely 'dialogue interpreting' or 'face-to-face interpreting'. These labels may not be useful for institutional-professional purposes, but they are helpful as an academic and pedagogical category in that they denote interpreting technique, i.e. 'interpreting a dialogue/conversation' and 'interpreting while you are directly facing your interlocutor(s)'.

Being a technique, dialogue or face-to-face interpreting can then be further sub-divided into a vast array of sub-categories according to where this form of interpreting is needed and implemented: in public institutions (especially legal/health/social), private institutions (international organizations, service sectors and the voluntary sector), schools, tourism, the media, business, and so on. Each of these categories denotes then a sub-area of application. We believe that this more general distinction is very useful as a pedagogical tool because in this way the trainer can teach interpreting skills *per se* and is free then to apply these skills, once the students have learned them, to a vast number of fields, especially those fields which are most relevant to the labour market in that particular country and to the students' own career choices.

To date then, taxonomies are far from homogenous, and as yet no definitive description has been offered to delineate this professional role, reflecting the simple fact that the applications of face-to-face interpreting are so wide-ranging and varied that any attempt to provide universal definitions would serve no purpose.

1.3.2 Community versus conference interpreting

We believe it is also important to make the distinction between community/public service interpreting and conference interpreting in the simultaneous mode. Even if the students will never actually be required to study or practise simultaneous interpreting, it is a useful distinction to make for pedagogical purposes. (At an interpreting school, this step will obviously be redundant.) Conference interpreting is a profession that has always captured the more vivid reaches of the students' imagination, conjuring up images of high-powered academics or politicians, high-tech booth equipment, the allure of working in international organizations such as the United Nations or EU, and the knowledge that the conference interpreter's job is a difficult one in which translation decisions are taken at the speed of lightning, in real time in an extremely complex neuro-linguistic process.

We remind the students that community interpreting takes place in institutions that generally cater to the public, rather than to the more 'closed circles' of the conference interpreter's professional milieu. The institutional aspect of the practice thus colours the understanding of the practitioners' role, both in their own minds and in those of the service providers. To position the various interpreting activities and professions in a wider perspective, we also remind the students that the social prestige of each field and institution (also tied to financial and political resources) reflects the attention they receive:

- science, academia, high-level politics (conference interpreting, prestigious training programmes);
- corporate business (dialogue interpreting, tailor-made training programmes);
- national health, immigration, social services, refugees, police and prisons (few specific training programmes at third level institutions).

It is useful to contextualize these professions in socio-economic and institutional terms. More important, however, is to describe the way simultaneous interpreting takes place and to compare it to face-to-face interpreting, making it easier for the students to understand the differences in mnemonic strategies and the neuro-linguistic processes through which they are implemented.

1.3.3   Sign language interpreting

One area of interpreting which has not yet been mentioned in this book, but which is fully established in many countries around the world, is interpreting for deaf people: sign language interpreting, which encompasses a range of different signed languages and language varieties (e.g. British Sign Language versus American Sign Language). Indeed, it could be argued that sign language interpreters began to develop professional organizations, standards of recruitment and ethics long before spoken language interpreters did, and even that the rise of community/public service interpreting as a discipline is due in part to the enormous amount of research and practical work done in sign language interpreting. It is, unfortunately, far beyond the scope of this book to provide material for trainers or even address this issue in any depth.

1.4   Creating a positive global effect by educating service providers

The focus of this book is the training of interpreters. Although this is by no means a novel idea, we would like to conclude this chapter with a brief comment on the need to raise awareness among service providers, to 'train' them to use interpreters to the best of their ability and potential, and also to be aware of the boundaries of the interpreter's role and tasks. In this context, by 'training' we mean sensitizing and raising awareness rather than just 'teaching'. This would help minimize the ambiguity and confusion that is a spin-off of a situation in which the interpreter's professional role – in many parts of the world – is still defined less by virtue of their own profession and professional standards and ethics than by those of the institutions they serve. In some countries, training courses on how to use interpreters in the public services are extremely popular amongst the service providers themselves (for example, in hospitals), but this has yet to become mainstream practice.

Awareness on the part of the institutions and primary interlocutors (doctors, social service workers, legal officials) of the problems inherent in interpreter-mediated communication greatly improves communication between the institution and the foreign speaker, which is a goal in itself. Furthermore, it is cost-effective both in the short and long run. This aspect, we believe, should be built into the course through class discussion. It is also a part of the general process of confidence-building and assertiveness-training for the students. The knowledge of how crucial their profession is to institutions and service providers should foster in the students an approach and manner through which they themselves transmit this message to those around them.

One aspect of this 're-education' process is also to focus on the positive aspects of the interpreter's role and the benefits they bring to the institution. Interpreters, students should be reminded, not only provide an essential service in a multilingual, globalized world, but they can play a positive and proactive part in establishing communication between interlocutors from different language communities and thus bring about a deeper, more robust and long-lasting mutual understanding and rapport between members of different linguistic, cultural, ethnic and corporate communities.

2
Theoretical Issues: The Impact of Cross-cultural Communication, Institutional Hierarchies and Professional Ethics on Interpreting

2.1 The impact of cultures on translating

Of all the issues addressed in studies of migration, globalization and intercultural communication (institutional and non-institutional), cross-cultural difference is perhaps the most important for our purposes. Culture is at the heart of the very processes of identity-building and personhood both at the individual and collective level. Indeed, the argument is almost circular: culture could be defined as that group behaviour which creates common behavioural patterns and serves to consolidate and define a group. At the more private, secluded, individual level too, a person's identity is still formed by the norms, expectations and customs of the group(s) in which that person is a part – by conforming or not conforming, by re-creating and challenging expectations in an infinite number of ways.

And if it is true that in any given, relatively homogenous, monolingual culture misunderstandings due to the imperfections or limitations of human language are rife, how much more rife must they be in a pluralistic monolingual metropolitan culture? Add to that the ethnically governed cultural differences and language differences inherent in the migration process and one might speak of the 'miracle' rather than the 'problem' of interlingual and intercultural communication. Clearly, a single chapter in a book such as this does not give us the scope to discuss the enormous range of issues that might be addressed in a study on intercultural communication, so we will limit ourselves to pointing out how cross-cultural differences might impact on the work of the interpreter.

Since this form of communication is almost always carried out in the workplace and in public institutions, we will be looking at institutional language rather than 'casual speech'. We will also be considering the workplace and the institution (or a combination of these) as a micro-culture unto itself (a 'corporate culture' in a very wide sense of the word), as yet another cultural sphere that must be negotiated by the various players in the game. The idea of 'corporate' or 'institutional' culture also reminds us of the fact that culture is far from being a monolithic, delineable and static entity that coincides neatly with an ethnic – or any other – group, but rather a plethora of 'smaller' and 'bigger' cultures that overlap and intertwine, co-exist and collide sometimes seemingly almost at random.

In much the same way, individuals build their private and group identity by drawing on this plethora of groups, by becoming 'members' of the groups by necessity and/or choice, overlapping between the various spheres of action, sometimes in peaceful co-existence and at other times in fraught tension. (Identity can be ascribed from birth – such as membership in a family – or achieved through work or other activities. The literature on these topics is vast, but see for example Trompenaars and Hampden-Turner 2000.) Thus, each individual's identity construction is a constantly changing, fluid process that must be negotiated and re-negotiated, not least when in contact with new members of other cultures (in the meeting with new migrants) or when moving to a new territory (when migrating to a new place). This applies also to the interpreter. (We would not want to over-emphasize this point either, however: at the end of the day each human being, anywhere in the world, is an individual – not only a collection of cultural imprints and defined by their socio-cultural context, but a unique person with a distinct personality from birth; indeed, this makes intercultural communication even more complex.)

In the interpreting scenario, with the need to communicate across this plethora of variegated cultures and identities – intensified by the fact that each of those cultures and identities is ethnically and territorially distinct – the players are attempting to negotiate those very identities without being able to draw upon that primeval communication tool that we take for granted every minute of our lives, namely the ability to speak together and to communicate ideas through speech.[1] Because our entire inner world (collective and individual) is channelled and manifested through verbal and non-verbal communication, this channel comprises our only means of gaining what we want or need (information, services, companionship, goods). And effective communication is vital for the physical, judicial or other well-being of one or both parties in the encounter. In our scenario, this absence of a common language is compensated for by a third participant who must draw upon their competence in each party's culture and language to allow them to understand each other as far as humanly possible. One does not need to resort to an absolute Whorf-Sapirian approach to appreciate the complexities involved for this third party: namely interpreting the complex interface of each of the interlocutors' cultures and individual personalities and re-creating these in a habit or form that does not necessarily contain corresponding expressions of the other party's words and concepts. And – as mentioned in chapter 1 – this must happen both at the terminological and linguistic level as well as at the pragmatic level for it to have the same pragmatic effect on both or all participants, as one might argue is the ideal objective of the translation process.[2] Add to this complex network the institutional, social and personal asymmetries of power and personality, including that of the interpreter, and the picture becomes very challenging indeed.

Fostering a critical self-awareness among students of the fact that their own views are by definition filtered or biased will help them put themselves in their interlocutors' shoes and will help them understand, express and bridge the communicative divergences between the parties without resorting to simplified stereotypes that might jeopardize the communication process.

We have claimed here that the rules of communication, as well as the language, customs and behaviour of every culture, are unique, but also that communication forms differ widely in public institutions in the same culture. Numerous scholars have suggested that hierarchy and power differentials are crucial to discourse forms in both Western and so-called traditional societies; the list is long, starting with linguistic anthropologists such as Foley (see 1998), Duranti (see 1997) or even the traditional works of early sociologists. Even ordinary conversation, which is, or is at least considered to be, 'reciprocal' and therefore largely 'egalitarian', is often strictly governed by power dynamics (familial, social and professional status, age, gender, kinship). This also applies to discourse in institutionalized settings (see Firth 1994; Sarangi and Roberts 1999). As in all hierarchical relationships, power asymmetries between participants in an interpreting context are manifested in a number of ways.

2.1.1 How to deal with hierarchical communication

Culture-specific communication modes that are affected by the positioning of the individual in society and in the institution/workplace, and especially the conception of hierarchy, require of the interpreter a skilful managing of discourse strategies such as turn-taking (where the interpreter is a speech coordinator, responsible for floor-management).

Turn-taking strategies require management of the length of speech chunks, interruptions, silences and body language. In order to respect the rules and norms of appropriate communication codes on both sides, the interpreter may have to function as a 'buffer', inserting or offsetting hedging strategies or (in)directness, and this is an extremely difficult balance because it may seem to clash with the interpreter's code of ethics (the tenet of 'accuracy'). What complicates matters further for the interpreter is that there is almost always an inherent power asymmetry, where the interlocutor with the power base (the representative of the institution) is in control of communication rules and has the upper hand, so to speak.[3] The interpreter is thus not entirely free to move between the communication and cultural codes without somehow paying homage (or possibly just lip-service) to the authority of the institution and/or its representative. The service provider's expectations (or indeed the institution's goal and regulations) might be at odds with the interpreter's self-appointed and self-perceived task (according to a professional code of ethics and the nature of the profession). The ideal situation would thus be the active collaboration and participation of all parties and a clear understanding of the interpreter's part in the process, as well as a clear understanding of a common objective and goal, rather than seeing the communicative act as a struggle for one-upmanship and fighting for the floor and/or one's own message.

2.1.2   Politeness management

A crucial factor in interpersonal relations in the professional as well as in the private sphere is the notion of participant roles (speech roles and social roles; this widely used terminology of speech participants hails from Goffman 1981) which are in large part negotiated through the use of culturally governed politeness strategies in order to establish, negotiate and/or clarify whatever happens to be the particular relationship between the interlocutors in that particular situation. 'Politeness', in a very wide sense of the word, is crucial in construing relationships but it is also extremely insidious in that it is so open to misunderstanding. (The debate on linguistic politeness has been flourishing since the work of Brown and Levinson [1987], and many of the early theories have been criticized, especially through a cross-cultural lens – see Bowe and Martin 2007; addressing these issues is simply beyond the scope of this chapter and not truly necessary to the aim of our book, which is essentially of a practical nature.)

In the reconstruction phase of the interpreting process – from hearing and comprehending to re-articulating – the interpreter must thus identify the pragmatic strategies through which politeness is construed and take these into account when attempting to recreate those same features pragmatically to the listener in forms that may vary greatly from the original utterance. Brown and Levinson's [1987] fundamental work of 'face', now critiqued for many of their simplistic and universalistic assumptions but none the less fundamental to linguistic disciplines, is an excellent starting point from which to describe to the students how such a basic communication feature as 'face' is so culture-bound. The British, Scandinavian or Japanese privacy norms, for example, require a strong focus on 'negative politeness' – that is, refraining from imposing or intruding on people's 'space', whilst less private cultures, such as the Mediterranean ones, will privilege positive politeness strategies that strengthen the interlocutors' self-image through compliments (Kate Fox's brilliant and humorous 2004 study *Watching the English. The Hidden Rules of English Behaviour* contains an excellent description of face negotiation among the English). Demonstrating to the students how easy it is to misunderstand these culturally ingrained and internalized practices cross-culturally, encourages them to process their own role in the translation of pragmatic conversation elements such as politeness and helps them find strategies to carry out this language–culture transition.

If we backtrack very briefly to the discussion on socio-cultural and institutional asymmetry in chapter 1, we see how essential it is to recognize the importance of cross-cultural awareness. If, as we have suggested, the communication format which most effectively gives access to essential services is that of the host country, then the outsider must be able to 'play by the rules', or be helped to 'play by the rules' in order to effectively access these services. In other words, if politeness competence is essential for communicative success – i.e. each party's success in obtaining the desired outcome of the exchange – then this will be accomplished best by staying within the dominant politeness system. Communicative competence thus hinges on a speaker's ability to 'perform' those codes that are most appropriate to the particular culture and the context (professional and/or social) and to use it to their advantage. Those conversation participants who do not speak the host country language or who are unfamiliar with the communicative and cultural codes, will always be at a disadvantage. As Fairclough says (1995: 1), unequal discoursal ability leads to an 'unequal capacity to control how texts are produced, distributed and consumed (and shaped)'. The consequences of this 'communicative asymmetry' can be very grave indeed: a client's indirectness or refusal to contradict or say no to the questioner could be interpreted by the judge or police officer as shiftiness or – worse – untruthfulness (whilst for the client it may be a politeness marker, a sign of respect). Communicative success (comprehension and participant response) depends, then, on the speakers acting out these codes cooperatively. Of course, communication can never be one-way, but will always depend on the other participants' comprehension and response. In her capacity as language expert, cultural informant and communication facilitator, the interpreter is the only person in a position to put the speakers on the same level.

The other factor that affects cross-cultural interpersonal relations through politeness strategies, then, is how individuals manage institutional hierarchies and communication with service providers – the doctor, the judge, the police officer, the teacher or headteacher, and so on. Culture-bound communication strategies are enormously important here: both the vision/understanding of the particular institutional (health, legal) system, as well as how the hierarchies within that system are governed by very specific strategies. Where in many Western countries, directness, upfront truth-sharing, direct eye-contact, a more directly egalitarian approach and other interpersonal strategies are considered to be markers of respect, other cultures prefer the opposite, as suggested above. Common across many parts of the non-Western world are strategies such as acquiescence (even if one is acquiescing to what is clearly not true and seemingly to one's own disadvantage), indirectness, never contradicting the interlocutor; avoiding eye-contact, avoiding excessive speech, and patronage of the person in the 'superior' position. In other words, positive face towards the interlocutor and personal face loss towards oneself (part and parcel of the same 'contract') may have higher value than establishing the objective chronological sequence of events. In many non-Western cultures, these may be more effective strategies in the furtherance of one's own situation than the dictum of 'telling the truth' which is, seemingly at least, a basic norm in much of the Western world. These strategies may be anathema to Western social organization and it is not hard to see how this can cause grave misunderstandings that may hinder access to justice.

2.1.3 Business settings

In the business setting, cross-cultural differences have become a recurrent theme. This is not surprising, as million-dollar contracts hinge on the ability to persuade the 'opposing' party to reach an agreement, and thus the need to appease and please. One interesting example we found of how this impacts on interpreting strategies comes from the well-known business management scholars Trompenaars and Hampden-Turner (1997: *Riding the Waves of Culture: Understanding Cultural Diversity in Business*), who claim that in Japan – a group-based collective society where the individual's identity is constructed through their membership in the group – the interpreter feels more strongly her sense of duty as a part of the group and the corporate team than as a professional interpreter who is completely neutral, impartial and in a position to make her own professional judgements without being conditioned by the needs or opinions of either group. Thus both her ethics and interpreting strategies change.

Although numerous more recent studies exist, for classroom use we find the models of Geert Hofstede (2001) and Fons Trompenaars and Charles Hampden-Turner (1997), used by a whole generation of business and communication scholars, both manageable enough and detailed enough to be helpful to students. Their models are similar in many ways: both identify features of cultural behaviour that they believe are common to certain countries; they have conducted wide-ranging and in-depth empirical studies in business settings the world over through which they attempt to place cultural values in a set of features identified a priori. We find that one of the most useful categories in explaining to students how culture affects communication is that of individual versus collective identity and group formation (alluded to above), following basic categories established by sociologists and anthropologists over the last century. The authors suggest that in collectivist, group-based cultures the interlocutors' objective during a business negotiation tends to be more holistic, aiming at creating a long-term relationship with their business partners that goes beyond the details of the negotiation at hand. Each of the actors is a participant in a socio-cultural event and the partners work together as a team; the company is a 'family', in a manner of speaking, where the behaviour of each member affects the other members, and thus harmony and negotiation are highly prized. The role of the businessman/woman is thus both social and professional. These roles are played out through communication strategies such as acquiescence and indirectness. In more individualist societies, however, the aim of partners in a negotiation tends to be more immediate, to reach an agreement or settlement quickly, without the need to create a more encompassing relationship (this may or may not happen, but is not an intrinsic part of the negotiating process). Social activities and relationships are appreciated, but outside the context of the 'meeting room'. Therefore, the way in which these professional values are communicated will be different in collectivist and individualist societies.

This representation of complex cultural processes is clearly an overly generalized one, as is our brief mention of Hofstede and Trompenaar's and Hampden-Turner's models, but the students may find it interesting to investigate these further. The models we have mentioned here relate specifically to the business setting, but the cultural values identified in them apply to most areas of institutional and private life and are therefore very helpful indeed in showing the students how value differences affect communication.

2.1.4   Greetings and non-verbal communication

We have claimed repeatedly that cross-cultural differences are embedded in and played out at many levels of human life and in many forms in interpersonal relationships. One of the forms in which they are played out, which is also related to the collectivist versus individualist identity formation we have just mentioned, is the use of greetings and leave-takings. These are an essential part of conversations of all kinds, in all contexts and at all levels of formality, and their manifestations are extremely culture-specific. In some cultures, greetings are minimal, a simple 'Hi' and nod of the head is considered sufficient. This is typical of so-called individualist and egalitarian cultures. Elaborate greetings asking about the health of the interlocutor and possibly the family are typical of more collectivist cultures. Greetings as conversation openers set the tone for what is about to take place and as conversation closers establish what has taken place. Greetings, like politeness (the two are clearly closely connected), are a useful gauge of social relations and must be used appropriately to achieve successful communication: they establish, reinforce and acknowledge interpersonal relations. They can foster both harmony and dissent (by creating distance and asymmetry). However, used as politeness markers they are (or the lack of them is) extremely susceptible to misunderstanding and may lead to tension and hostility in intercultural communication. Greetings are often accompanied by very precise non-linguistic markers such as hand-shakes, a nod of the head, a bow, a smile or lack of smile, eye-contact or lack of eye-contact, body-direction, etc., and these too are susceptible to cultural misunderstandings.

Students should be taught that the interpreter must be able to read and reproduce – in a manner appropriate to the interlocutor(s) and to the situation – these conversation openers and closers that are so crucial in establishing interpersonal relations, and that they may be decisive for the outcome of, for example, a business negotiation. Even a cursory library or internet search will show that much research has been invested in intercultural business studies on how to use these inter-personal markers appropriately – precisely for this reason. (Trainers must of course find examples that are appropriate to their needs and language pairs. Some examples can be found in the dialogues in chapter 6.)

Interpersonal relationships and politeness strategies can, as men-tioned, be expressed non-verbally at a more general level. For example, how close you can stand/sit to another person(s) is culturally defined, as is how directly you can look at them, how loudly or directly you can talk to them, and not least whether or not and how you can touch them – these practices are all expressed in ways that are more or less rigidly codified by each society. These issues are important not only because an interpreter must in some way manage that in-between, undefined space between the interlocutors, but she must know and proactively decide how to act upon it: to interrupt an interlocutor by touching them on the forearm or shoulder can communicate reassur-ance and respect in one culture, whilst this same act might signal for another interlocutor intrusion and lack of respect.

2.2   Bridging cross-cultural differences: intercultural transfer competence

By *intercultural transfer competence*, we mean firstly the ability to identify those salient verbal and non-verbal communicative features that the interlocutors unconsciously rely on and consciously employ to com-municate. Secondly, we mean the ability to translate these features into the interlocutors' language so that they perceive as much as possible of the assumed intentions of the speakers.[4]

What kinds of competencies do interpreters need to recognize cul-tural differences, and what kinds of strategies do they need to deal with them? At the micro-level, the interpreter will first and foremost need a good command of both languages, but also field-specific terminol-ogy and language varieties of lingua francas (these varieties exist, for example, for English, French, Spanish, Arabic and Chinese) and ide-ally, also cultural aspects relating to each of these varieties, as well as familiarity with regional variations, such as geographical accents and sociolects. As well as language proficiency, the interpreter will also need cultural competence and proficiency in both/all cultures (to the extent that this is possible)[5] and the ability to transfer that knowledge to the other party ('transfer skills'). These two competencies (which are not of course realistically delineable into two neat 'skills') are impor-tant and difficult.

In this book we will not be recommending any specific 'cultural competence model' (there are many such models available in the literature and through the internet);[6] but we suggest that if the trainer is able to convey a sense of curiosity and awareness in the students of how human beings are socialized from infancy into a particular model of behaviour and communication, and convey the extent to which these cultural systems differ – perhaps having little or no overlap, much of the groundwork will already have been laid. In other words, we are suggesting that the very awareness of, and sensitivity and openness to these issues will automatically activate certain communication, or repair, skills in the students when meeting interlocutors from different cultures. This awareness will also lead to self-reflection regarding their own culture-embeddedness and how their own communication forms are informed by their own culture(s). An awareness of their own cultural positioning and subjectivity should also lead them to the knowledge that their own prejudices and stereotypes may affect the interpersonal aspects of interpreting performance. Equally important, furthermore, is the awareness that each person is an individual in a larger group, a product of idiosyncratic traits but also their own, personalized, individual journey through life. If a general pedagogical objective is to provide the students with an awareness of culture-specific communication systems and of their own place within and between these systems, the ultimate aim must nevertheless be to provide them with the 'cultural production skills' to translate these culturally specific elements, verbal and non-verbal, into a meaningful communicative act that is constructive for each of the interlocutors.

In other words, it may not be difficult to identify (for the interpreter and for the researcher) cross-cultural differences in the communication systems at issue, but what the interpreter chooses to do with them is quite another matter. All of these communicative norms that vary so widely must in some fashion be negotiated by the interpreter: she may choose to ignore them and simply translate the verbal/lexical component of the message, or she may choose to communicate the cultural, social and pragmatic weight of the message in a form that is comprehensible to the listener, to the extent that this is possible. Or rather, given the relative incommensurability of communicative systems and languages, she may make a brave attempt to bring together disparate communicative systems and disparate actors to put them in a position to understand each other, again to the degree that that is possible, and to the best of her ability. The range of strategies from which to choose is limited, but the consequence of each is potentially significant in its impact. The interpreter's responsibility is indeed daunting: it is precisely in the ability to recreate understanding between the parties in which the real challenge for the interpreter lies.

In order to develop intercultural competence in our students we need to train them to recognize and understand cultural differences on both sides, to increase intercultural awareness, including the subjectivity of their own cultural perceptions, and to know how to employ intercultural communication skills that will make this communicative act possible. Let us see in more practical terms how this can be done.

2.2.1   Coordinating discourse: cross-cultural variation in conversation management

One of the most important interpersonal communication issues for interpreters and thus potential problems in interpreting is floor-management, more specifically co-ordinating the length of chunks used by the speakers to allow the interpreter to remember all the salient elements in any chunk of speech and not jeopardize accuracy. Indeed, it is absolutely crucial that students gradually learn to control the floor and the speech-situation to be able to perform according to their professional mandate – that is to translate fully and accurately what is being said. To do this, they must not only recognize and reproduce those social codes embedded in culture that we have discussed above, but at a more practical level they must learn to interrupt when necessary and to recognize their own mnemonic limitations: to know how long/complex a chunk of speech they can remember and to interrupt when this limit is exceeded. Needless to say, students' mnemonic capacities will be very individual and they must learn to respect their own limits.

Our aim is to train students to know exactly when to stop the speaker, allowing them to finish a chunk of speech that is meaningful but not long enough to forget what has been said. To do this, it is important that the interpreter finds a natural point of entry without disrupting the speaker unnecessarily, but without listening to more than she is able to remember to give a full rendition of the utterance. The issue of accuracy is absolutely crucial (especially if note-taking is not allowed). Although it is a 'technical' issue and in theory easily solved, it is far from easy in practice: interrupting a speaker in a classroom situation (either co-student or trainer) can be difficult, but in real-life interpreting situations, especially at the beginning of their career, this of course can be even more difficult, complicated by social, professional, age- and gender-related hierarchies. Students need to be aware of how intimidating these interpersonal relations can be in institutional contexts and how they can hinder responsible conversation management.

The spatial and acoustic logistics of the room can also be difficult: a small room with three people sitting close together within easy hearing and sight distance (and able to resort to body language to convey meaning) is very different from a large courtroom full of people, with poor acoustics and an uncollaborative judge. In addition to the practical problems, an interpreter needs a great deal of self-confidence to challenge the authority of the court and interrupt the speakers. The training video *Points of Departure* (see details on page 236), which we have found very useful, has a good sketch illustrating precisely this point. Having practised this in class – not just 'knowing it' but actually having done it – will give the students the know-how and the confidence to put it into practice in 'real life'.

This is also a good opportunity to discuss at a more general level the culture-specificity of conversational cues (e.g. the start and finish of turns signalled by intonation cues, voice level, pauses, body language, physical proximity, eye-contact, discourse markers, the lengths of gaps and silences, the threshold for overlapping speech). Seemingly incongruous speech functions – questions in the declarative, requests in the interrogative, and so on, as well as irony, humour and understatement – should be discussed, ideally using these features in role play. The level of tolerance for overlapping speech and/or silence is another aspect that impacts a great deal on turn-taking and the ability of each speaker to take the floor. (For speakers from cultures with a low tolerance for overlapping and a higher tolerance for silence, who keep waiting for a suitable entry point, it is sometimes difficult to get a word in edgeways!) To illustrate this with an Italian example, we like to show the students recordings from a well-known political talk show where the level of overlapping speech, especially during election time, is so high that it is often difficult to make sense of the speakers' words. We sometimes show other excerpts from natural speech on television to make students more aware of turn-taking traditions in their own country.

In short, the interpreter must master the task of combining verbal and non-verbal cues and of reproducing them appropriately. As with all speakers engaged in natural discourse, she must have the ability to recognize and exploit rhythm, tone and intonation in order to identify chunks/signals/turns and to identify irony, sarcasm, respect, emotion/ affect, humour, aggression, fear, approval, disapproval, formality-distance, camaraderie. In the final analysis, however, a successful outcome depends on the good-will and collaboration of all parties.

2.2.2   Coordinating talk: coordinating emotional talk

There are some situations in which coordinating talk can be extremely difficult, and that is when intense emotions or states of mind are involved, for example in particularly distressing legal settings such as the Hague Tribunal and other violent crime situations, UN and Red Cross fieldwork, war, the emergency ward, terminally ill patients, as well as much police work. The interpreter needs to learn first of all to distance herself, as far as is humanly possible, from the interlocutors and the situation at hand and to process her own feelings in order not to risk serious burn-out (see Baistow 1999 on vicarious trauma), but also to be able to interpret effectively. There are rather obvious strategies to follow here, such as not interrupting, taking notes so as to not have to ask for repetition (discreetly if note-taking could be seen as threatening), using unthreatening body language and being cautious with body-proximity and touch (i.e. respecting the culturally appropriate norms). However, the interpreter must also be aware of how emotions affect the flow of conversation and be alert to the fact that emotions are signals and cues of important information and cognitive content (see Kleinman and Copp 1993).

2.2.3   When the interpreter is drawn into the conversation

To make things worse, the interpreter is often interrupted by the speakers. Particularly in moments of tension, the speakers may turn to the interpreter – who until then has been ignored as if she were completely invisible – for assistance: 'Explain to x that that simply isn't the way things are done!' At that point it becomes very difficult for the interpreter not only to coordinate the floor but to extricate herself from active participation in the conversation. It is a good idea to discuss with the class what to do in these situations, which are of course connected to professional ethics. The solution may lie in a compromise and will also depend on the understanding of the role of the interpreter in that particular country and in that particular situation. This requires skill, practice and self-confidence and it is useful to insert situations like these into the role plays.

2.3   The code of ethics: the interpreter's role, responsibility and tasks

2.3.1   Knowledge and competence

The required skills in order to maximise accuracy can be listed as follows.

Knowledge
- **Adequate language comprehension**: adequate comprehension of textual and pragmatic features.

- **Comprehension of non-verbal features:** an interpreter must be able to read a speaker's individual non-verbal language (hesitancy, embarrassment, fear, politeness, taboos, understatement, silence, repetition), which may signal important non-verbal messages that are crucial for communication.
- **Field-specific and institutional knowledge:** an interpreter must be familiar with institutional terminology and have a background knowledge of the health system, legal system, administrative system – and the differences in the systems of the target and receiving cultures.
- **Cross-cultural awareness:** an interpreter must be acutely aware of any cross-cultural issue that may lead to communication breaches and must therefore be intimately versed in the cultural and communication norms of both the client's and the service provider's cultures.

Competence

- **Active language ability:** the ability to produce meaningful speech in L1 and L2 that fulfils the interlocutors' communicative requirements.
- **Accuracy:** the ability to translate the speaker's utterance as fully and accurately as possible.
- **Transfer:** the ability to reproduce and transfer the various speakers' utterances by drawing on the above competences – cultural, interpersonal, pragmatic and linguistic.

Issues of knowledge and competence are discussed more fully in chapter 4.

2.3.2 Attitude

According to most codes of ethics, impartiality (distance) and the empathetic rapport with client (bonding) are the key aspects of appropriate interpreter attitude.

Impartiality

An interpreter must avoid over-identification with the institution. This will always imply not adopting an overtly judgemental attitude towards the interlocutors. The question of *advocacy* versus *impartiality* is a burning topic in community and other forms of dialogue interpreting, and it is a complex issue. Although in theory most practitioners and researchers would agree that interpreters should never become 'involved' in the interpreting process, this may be difficult in practice. Nevertheless, we believe that the important issue here is that any emotional or other involvement should not overtly influence the interpreting process – to the degree that this is humanly possible. Even if the interpreter's professional approach is that of 'impartiality', her own experience, opinions, ideology and culture will, unknowingly, shine through her performance in many subtle ways – facial expressions, gestures, tone of voice, politeness level, etc., which are extremely hard to govern. Rather than attempting to censor these interpersonal elements and aiming for a 'robot-like' performance, the interpreter should be trained to be aware of them and use them constructively in building the bridge between client and service provider, not by becoming an 'advocate' of either side but by being a fully participatory and aware interlocutor possessing essential information that both parties need in order to communicate effectively.

Bonding

Frequently, a client in a vulnerable situation – perhaps a person who arrived in the host country recently and has little or no contact with fellow countrymen, who may be confused and insecure about the customs and communication forms of the host country, and who feels vulnerable and suspicious about real or perceived hostility – will turn to the interpreter as a friend in need, a point of contact between his own life, language, world and the situation in which he presently finds himself. At the same time, the host country representative (doctor, police officer, immigration officer, social service worker) might see the interpreter as a mouthpiece and representative of the host institution.

Bonding, feeling sympathy or empathy for and identification with someone, is a common and perhaps inevitable phenomenon in this form of interpreting for a client who is in a vulnerable situation, not only because of their inability to understand the host country language but because they are usually facing some form of crisis or lack. The client will frequently reach out to the interpreter because of this vulnerability and because she is the only person who understands his language. For these reasons, bonding often begins with the client, but the interpreter might also identify with the client for other reasons – cultural identification, sympathy, and so on. Bonding between parties is a natural process, especially when the client and interpreter come from the same ethnic group, if that group is a small minority, and especially if it is a persecuted minority. Indeed, positive bonding is a way of constructing a healthy working relationship between the three parties, putting the clients at their ease, facilitating the working relationship between client and institution, and it is furthermore a way of building general trust. However, bonding should never be allowed to interfere overtly in the communication process, as it may interfere in the interpretation and may jeopardise impartiality, putting even more psychological pressure on the interpreter or the other parties.

Indeed, one of the most difficult challenges for an interpreter is that of knowing exactly what her role is, where to draw the limits or set the boundary between her own professional role and the desire to help as a human being. If the interpreter clarifies to the interlocutors what her role or mandate is during a pre-session briefing, for example, this could help considerably.

Strong nerves

An interpreter must therefore have strong nerves and an awareness of her own professional role, responsibilities and limitations. She will not infrequently have to confront sensitive and painful issues both in actual verbal communication and the context in which such communications are situated: violence and aggression, at the police station; terminal disease, shock, violence, death, mental illness, at the hospital or clinic; issues of violence, child abuse, rape, torture and war in refugee applications: emotional and psychological stress, conflict, emotion, anger, fear are all issues an interpreter may have to cope with.

2.3.3   Conduct

The interpreter should be careful about how she uses the information she is privy to.

Asking the interpreter for advice

It may also happen that the client asks the interpreter for his or her opinion or advice (there is a good sketch illustrating this issue in *Points of Departure*). When this happens, the interpreter may provide or re-state information that will assist the client in making his or her own decision, but should not give her own opinion with the aim of influencing the client. In this case too, the interpreter can explain that giving advice would mean breaking her code of ethics.

To avoid seeming to be offering opinions, or to avoid seeming to be taking sides, interpreters should avoid discussing the case when left alone with the client in the waiting room and all interpreters should avoid having 'side-conversations'. Such side-conversations, although unavoidable sometimes, may give the impression that the interpreter is not sharing all information equally with all parties or that she is taking sides. It is sometimes necessary to clarify issues with either the client or service provider during the interpreting session, and one way of managing this is simply to explain to the other party that you are clarifying linguistic or cultural issues.

One good anecdote to use with the students to illustrate this is the following:

> The doctor asks the patient a question. The interpreter and the patient get into a long discussion, while the doctor sits and waits, completely left out. Finally the interpreter turns to the doctor and says 'She said no.' When the doctor asks exactly what the patient said, the interpreter smiles and says, 'Oh, it wasn't important. She just means no.'
>
> (adapted from www.xculture.org)

The interpreter must be very clear about the limits and responsibilities of her professional role. Sometimes the difference between giving advice and offering information on cultural differences or about institutions can be difficult to establish. Giving opinions or advice about the progress of a case to console or encourage the client can be tempting, but is highly inadvisable. Nevertheless, when an interpreter realizes that a misunderstanding is about to arise, for linguistic or cultural reasons, she should nevertheless speak up and take control of the situation to prevent or clarify any such misunderstanding. It takes a good deal of maturity and self-confidence to be able to do this, however – maturity which is acquired through training and experience.

Confidentiality

Essential to any interpreter's code of ethics is *confidentiality*. It is crucial that a client be able to trust the interpreter not to divulge any of the information that emerges during the interpreting session, especially to members of the client's own community. An interpreter can initiate the session by establishing her role and the limits of that role and by advising all parties in the interpreting session to refrain from saying anything they do not wish to be interpreted. Far too often service providers do not inform interpreters of this responsibility – both in the legal and health sectors. In legal interpreting, apart from the effects on the client, breach of confidentiality may affect the investigation and the outcome of the case; it may even put the client's life in danger.

So it is crucial that the service providers here take their part of the responsibility by informing the interpreter of the confidentiality requirement. It is a curious fact also that the perception of confidentiality may vary, in the sense that service providers may expect the interpreter not to divulge information about the case/client to outsiders, but will expect the interpreter to tell *them* anything they think is relevant or necessary for therapeutic/ investigative purposes, as if confidentiality were 'unilateral'. For example, if the interpreter knows from a previous interpreting session that a patient is omitting to provide information that might affect the therapeutic outcome, the decision becomes difficult. If the interpreter knows that the client is lying about welfare benefits or employment, it is perhaps less difficult: patients and clients have a right to decide what to share, with whom to share it, and – if they choose – to lie. What if the interpreter knows that her refugee client is lying about political conditions in the home country – does she tell the therapist/immigration officer? Making choices in a life-threatening situation where the information the interpreter possesses can save the patient's life is not difficult, but, as Pollard says (1997–1998: 31), it is those grey areas, those 'in-between' decisions that are hard to make. Each interpreter must make that decision by herself, according to the specific circumstances. These decisions will be easier to make if the interpreter has a clear but also a realistic understanding of her own professional role. It might also be useful for an interpreter to discuss such dilemmas with an experienced colleague/supervisor.

Practical steps

One way of making it easier for the interpreter to put into practice these guidelines is to discuss the case and/or problems with the service provider before or after the meeting. *Briefing* can help both the service provider and interpreter to avoid potential misunderstandings. Taking some minutes before the session to allow the interpreter and service provider to discuss terminological, technical or ethical aspects of the case, as is common – indeed required – in many countries, can be very useful. If the interpreter feels that misunderstandings have arisen during the encounter (especially terminological or cultural), she should ask to have a debriefing with the service provider to discuss these issues. This is an important and healthy process because it allows the interpreter to do the job conscientiously, but it also trains the service providers to appreciate the difficulties and skills of the interpreter, and also raises their awareness of the difficulties of cross-cultural communication – in other words, it is a way of sensitizing, if not actually training, the service providers.

An alternative to meeting before the session, can, as mentioned, be meeting afterwards in a *debriefing* session to discuss any problems that may have arisen, issues that are still unclear or may have led to a misunderstanding. It is also an opportunity for the interpreter to discuss difficult ethical issues or dilemmas with a doctor, psychiatrist or judge if she needs support, advice or simply empathy. In mental health and refugee interpreting (sometimes in police interpreting), this is particularly important, as the issues dealt with can be very emotionally and psychologically trying for the interpreter. Ultimately, decisions about whether to brief/de-brief, whether or not one can offer practical advice to the client, whether or not the interpreter should be allowed to be alone with client, etc., are, of course, not the responsibility of the interpreter alone but of the institutions themselves, and yet it is often the interpreter who finds herself expected and obliged to take such decisions and who must often shoulder the blame when communication breaks down.

2.3.4 Personal responsibility: acceptance of assignments

Conflicts of interest, which should always be disclosed, can arise for numerous reasons: an interest in the outcome of the situation, preconceived notions/prejudices, emotional reactions to the people/issues involved, monetary interest, lack of linguistic or cultural competence, background knowledge of, or friendship with, the client. Ideally, an interpreter should turn down an assignment when any of these conditions are present. In practice however, an interpreter will sometimes accept an assignment for which she does not feel fully qualified (linguistically or otherwise) because it is an emergency or because there are no other interpreters available for that particular language combination. To avoid liability, in these cases the interpreter should always make the situation clear to the people involved. As a rule, however, if the interpreter realizes that her level of competence is inadequate, she should not accept the assignment.

3
Background: Health Services, Legal Institutions and the Business Sector

In this chapter we will look a little more closely at *what* to teach in the three interpreting strands we have chosen to focus on and will explore the contents of each of these modules. Trainers can use the discursive material in this chapter as background material for their classes. In chapters 4 and 5 we will look more closely at *how* this material is to be taught.

First of all, a few words on *motivation*. We have already drawn attention to the fact that in many countries today interpreting services in public and private institutions are poorly organized and that this is a potentially serious obstacle to the well-being of millions of people. Fortunately, this situation is being addressed in many countries around the world, but others are still far behind. Local, contextualized information about the situation in their own countries is extremely useful for students, not least to be able to plan their own professional development. Students should know how language services actually function on the ground in their own country. They should be aware of logistical aspects such as where the main hospitals, police stations, courts, schools, job centres, benefit offices, trade-fairs in their towns/cities are and the languages that are most frequently used, how the interpreting staff are recruited, training facilities for interpreters in their cities/ region, and so forth.

However, the basic, 'bottom-line' parameter that needs to be addressed from the very first lesson and that will give the rest of the course its *raison d'être* is that students should be made aware of how directly (in)effective language services and (in)adequate communication impact on the quality of healthcare, legal assistance, the delivery of education, and so on, and that poor communication in these areas can lead to misunderstandings, tension, misdiagnoses, mis-trials, and other problems.

Interpreters play a major role as intermediaries between doctors and foreign patients and thus impact on professional liability in the health sector. The knowledge that good communication allows migrants, businessmen, tourists or foreign students access to all these essential facilities and saves institutions and the community at large both money and unnecessary effort and tension is highly motivating for the students, as is the knowledge that poor communication (monolingual or bilingual) may lead to further expenditure to repair earlier damage and to organize new interpreting sessions. If we as trainers are able to demonstrate to the students the difference qualified interpreters can make in improving communication and saving the hospitals, police forces, various offices and schools from unnecessary costs, and ensuring good quality services, this will also give them the knowledge that they too can make a difference in their communities.

Like most jobs, interpreting assignments require *preparation* and an interpreter should of course be well-prepared by studying for the assignment, unless she already has gained sufficient experience in the field. Like conference interpreters, preparing for an assignment (especially the first few times) will in most settings entail reviewing or updating terminology, but it also goes beyond terminological know-how. Each sector will have specific requirements, but generally speaking, interpreters should master cultural as well as institutional differences (including all the professional titles, positions and hierarchies and the appropriate forms of address). She should know which strategies to use to compensate for cultural or terminological gaps: paraphrase, explain, create neologisms, etc. We will address this issue more specifically for each of the three settings.

What features do they have in common?

Teaching interpreting in different settings naturally requires that the specificity of each setting is addressed separately, but in this chapter we have tried to group the commonalities of each of the settings by categorizing the features they have in common through their *structure,* which refers to the organization of the system, followed by the field-specific *contents* of the sector at issue.

When discussing the structure of the organization, the trainer needs to look at how its basic infrastructure and organization works and the people involved in this system, but also at its communication structure and how interpreters function within the system. This will include local, practical aspects such as how they are recruited, what their role is, how they interact with the institution, how much they are paid, practical-logistical aspects, and so forth. We have decided to categorize these as *organization/structure, participants* and *communication structure and strategies* (which include the emotional context as well as the interpersonal rapport and power asymmetries between participants). It is also important for the future interpreters to be aware of how the interpreting services themselves are organized. These aspects are of course deeply embedded in each other, but we will address them separately to give the trainer a more structured training tool that can be applied and adapted to each unique circumstance.

Secondly, the *contents* of each specific sector can be probed by trainers through field-specific *terminology, cross-cultural aspects* and sector-specific *ethics*. We have also included sector-specific *preparation* and sector-specific *briefing* in terms of the type of information necessary for both parties, and some sector-specific suggestions for further reading. The sections which follow will also be addressing the *mode(s)* that are most frequently used and most appropriate to each of these sectors.

A useful schema for structuring this part of the course would then be:

1. **Structure**
   - Organization and structure of each setting
   - Participants
   - Communication structure and strategies (including social and emotional context, interpersonal rapport and power asymmetries)
   - How the interpreting services are organized
2. **Sector-specific contents**
   - Terminology
   - Cross-cultural aspects
   - Ethics
   - Texts
   - Preparation
   - Briefing (specific information necessary for both parties)

In this chapter, we feel that the best way to illustrate these course contents is to use general notions; the structure of the organization/setting is so localized and country-specific that we can only give a few examples from our own situation in Italy to illustrate the basic schema. We hope this will provide the reader with a general teaching framework that can then be adapted to suit their own specific, local needs.

3.1 Interpreting for the health services

3.1.1 Organization and structure

Although the basic organization of the health sector and the health system is common to most Western countries, there are nevertheless a number of differences that are worth noting and that future health interpreters should be familiar with.

Whether or not the health system is public or private will have a significant impact on the practical organization of booking and paying for examinations and treatment, but will also affect the rapport between users and service providers. In a private system such as that in the USA, the patients will be seen more as 'clients' who are paying for a service and thus have specific rights, a situation affecting also the interpersonal rapport between clinician and patient, the way in which medical information is offered and the notion of the patient's choice of various treatment options (i.e. the patient is paying for a service and as a client/customer should have the widest possible range of choice; the patient is the final decision-maker). Many Americans are socialized to be active participants in all discourse situations, and to be equal participants in all conversations regardless of professional or social hierarchies. This is also typical both of egalitarian-style and individualist societies, where the individual's right to choose is held to be sacrosanct. In national health systems where the patient is seen as a tax-payer but not as a client, the rapport may be more marked by the staff being accountable primarily to their institution, community and/or country rather than primarily to the 'clients'; here, the medical professional is the expert who is offering not just advice but pronouncing 'doctor's orders' that they expect to be followed. In many collectivist societies, the asymmetrical rapport between health professional and patient is indeed encouraged rather than negated (as it would be in highly individualist societies), the patient expects to be told what to do by the medical professional. Because patients have such faith in the professional, they expect tangible results (often miracles!) and are therefore willing to pay the possibly deferential respect that they feel is appropriate to the expert's task and role. In many Western countries today there is a mixed public-private system; although in those cases the above description should be somewhat qualified, a country which has a long tradition of a national health system will, we believe, have a different basic rapport with the community than a country that has always had a private health system. (In very poor non-Western countries with no public health systems the rapport will be different again.) Lastly, in private health systems, insurance questions are crucial for interpreters who are often asked to translate or sight-translate (i.e. translate aloud in real time directly from a written text) forms and other relevant documents.

Practical issues for the newcomer

There are a number of more practical issues that affect the local organization of health services, not least the use of and the role of the GP (general practitioner, or the 'family doctor'). If and when the system of basic health services is different, a person coming to a new country will have to find information about how the new system works, and this is often difficult, time-consuming and confusing. Although the system may not be complicated for people who live there, it may be so for a newcomer for whom everything seems unfamiliar and who does not speak the language: registering with a local GP, queuing up at the GP's surgery (do you ask for an appointment or simply turn up?), explaining your symptoms, asking questions, knowing what to ask for, asking for a referral to a specialist (the patient may assume that the GP will be able to deal with all medical problems and procedures), booking the specialist examination (often at a third venue, such as the pharmacy or hospital), dealing with the logistics of getting to the specialist and possible pre-examination procedures (fasting before a blood test, urine samples), going to retrieve the results of the test, going back to the GP with the results, following instructions for treatment, knowing which medications are covered by the national health system, and so on. It is indeed a labyrinth, and many migrants to the West (we use the example of Italy) feel more comfortable going directly to the emergency ward in the hospital: the hospital is easy to find and you only need to go once, however long the queue is. The doctors at the hospital will fulfil many of the functions that are otherwise delegated to various actors in the health system: they examine you, take tests, diagnose and prescribe treatment. This direct route is used especially by families with small children. Also, hospitals may not be required to check residence permits.

Apart from the practical-logistical procedural aspects, there are many other variations in the health systems in Western countries, for example the use of local clinics versus hospitals, home visits by doctors, district nurses or health visitors, local paediatric clinics or health centres. Prenatal, infant and child care are organized very differently from country to country and in some countries doctors or health clinics expect mothers/parents to bring their children regularly for check-ups rather than just when the need arises. In some countries, children have access to school doctors and dentists, which takes much of the pressure off the parents.

Preventative and community medicine – vaccinations, PAP tests/cervical smear tests and mammograms, and so on – are offered through systems of regular medical procedures which vary from country to country. New migrants may struggle with the local system and may simply ignore what is on offer.

In hospitals, the tasks that are assigned to doctors or nurses are very different from country to country, as is the rapport between them (leading to significant cross-cultural communication differences that should be addressed by interpreter trainers). Wards and internal laboratories (for blood tests and other tests) are organized very differently, as are the internal bureaucracy and administration, and admission and discharge procedures.

Sexual health is also addressed very differently from country to country (and will to some extent be governed by traditional and/or political attitudes towards sexual practice). Sexual health and hygiene, contraception, sexually transmitted disease (especially but not only AIDS) will all be affected by the prevailing local attitudes towards such delicate issues. This will in its turn affect how such information is divulged to the public, the privacy of access to these clinics and aspects such as outreach work.

How the interpreting services are organized

Because the systems are so different from country to country, we will not be dwelling on specifics. Suffice it to say that students should be familiar with their own local situation and how interpreters in their country and communities are trained (if at all), recruited, hired and paid.

It is helpful for the students to be informed (warned!) of how interpreters are generally treated by the medical staff, what their tasks are (if these include written translation and possibly other, far more mundane, tasks such as accompanying the patients and their families, providing information, etc.). This is also a good opportunity to discuss the ethics of these 'extra tasks' vis-à-vis a more general code of practice for the interpreting community. Administrative and bureaucratic, as well as legal, aspects can also be discussed at this stage as well as practical aspects such as the possibility of pursuing a career in this sector.

3.1.2 Sector-specific contents

Terminology: becoming familiar with a variety of settings, subject matter and terminology

In all sectors we find a variety of different language terms and registers, so even when interpreters are fortunate enough to be able to specialize in one or two sectors, there is still a range of sub-languages to be learned and to be updated in order to keep abreast of developments in law, technology, politics, international relations, immigration and trade statistics, and so forth. This is perhaps even more true for the medical sector than for other sectors because it covers a particularly wide range of highly technical language(s), as Tebble (1999) notes, from emergency medicine to geriatrics to psychiatry. We have seen that interpreters in the health sector are also required to work in a range of different settings: hospital wards and clinics, consulting rooms, patients' homes and community health centres. They interpret not only for patients and doctors, but also for nurses, physiotherapists and speech therapists, occasionally for administrative staff too. They will often be asked to translate documents or to write documents and letters in the foreign language, or to contact foreign insurance companies and embassies over the telephone. Furthermore, as we mentioned above, health interpreters are often expected, albeit indirectly, to fulfil a range of 'non-professional' tasks, for example to function as psychological support and care-givers to foreign patients. This is dubious practice on the institution's part because the institutional role of interpreter or interpreter/translator becomes extremely ambiguous; but it is more understandable from the patient's point of view: in a particularly vulnerable moment in his life, the patient naturally clings on to the only point of contact between his own 'world' and the foreign institution. It is clear that the skills and activities required of the health interpreter are wide-ranging and thus require a certain amount of flexibility. In practice, interpreters tend to be extraordinarily versatile and adaptable, due perhaps to a combination of their training, personal experience and/or bilingual and bi-cultural backgrounds.

Register variation

Apart from the challenge of terminological variation, the range of registers in medical discourse also varies greatly. In one single medical interpreting session (for example a paediatric consultation) this might range from small talk with the parent, to baby talk with the child, then to technical terminology with hospital staff and then with the parent in explaining what has been said. We have provided an example of a systematic ordering of discourse variation (suggested by Tebble 1999) in chapter 4 which deals specifically with medical discourse but which can be applied, albeit cautiously, to all the fields we are discussing here. This model is a useful starting point from which the students can reproduce distinct discourse patterns in their own renditions. We must remember however that these 'neat' models do not reflect the complexity of real-life conversations, with the time-pressure, complicated logistics and human emotions in real-life workplaces, especially in the chaotic and pressurized world of hospitals.

Communication structure and strategies

Similarly to interpreting in the legal sector and certain other forms of community interpreting (as opposed to business interpreting, interpreting in schools, interpreting for the media and conference interpreting), the interpersonal rapport in a health service setting is marked not only by institutional – and often social – hierarchies, but even more importantly by a very loaded emotional context. By definition, in a health interpreting situation the user is – to a greater or lesser degree – in a vulnerable physical and/or mental condition and therefore in an emotionally charged context. This emotional context will almost always affect the exchange of information, we believe, and should be taken into account by service providers but also by interpreters. Although it is clearly not the interpreter's mandate to provide the patients/users with the emotional support that they should be getting from the institution, the perception of their mandate does affect the communication at a linguistic level – the format of questions asked and answered, the information offered voluntarily or withheld, or just suggested and hinted at, the tone of voice that may be conveying important cues, the body language, and so on – as mentioned in chapter 2.

Aims of communication

In contrast to the conflictual and often hostile legal setting, in the health sector the primary aim of all parties is to exchange sufficient information to provide, if possible, a diagnosis and treatment, and therefore the talk will be (ideally) collaborative, aiming for a common goal, rather than being antagonistic and secretive. The aim, at least for the service provider, is to exchange information as fully and openly as possible without wasting too much time, and it is therefore easier, in theory, to establish a collaborative rapport than in the legal setting. This deeply affects the mode of communication and the exchange of information and will also usually lead to a more flexible approach where adding essential information *may* be acceptable (which does not mean that semantic accuracy is any less important – on the contrary, accuracy is vital for diagnosing symptoms).

In chapter 2 we discussed how verbal and non-verbal communication patterns and norms change radically from culture to culture and language to language, and we will not repeat this information here; but there are some practical issues that could be mentioned, such as the non-verbal aspects of communication, which are particularly important in a situation whose primary focus is the body and the dysfunctions of the body. Eye-contact, as Bischoff and Loutan (1999) note, is a useful communication tool when patient and doctor cannot speak directly, but in some cultures direct eye-contact is considered a breach of social norms.[1] Observing body language and tone of voice, facial expressions, gestures and movements can also provide the interlocutor with a great deal of information (although she must tread carefully here as much body language is clearly culturally constructed), and this information must in some way be conveyed to the participants (either by re-enacting it in a culturally meaningful manner, or by describing the behaviour if it is marked). The doctor has plenty of opportunity to observe the patient as the interpreter is speaking, especially if they are seated in a manner that favours such contact, and oftentimes this may indeed suffice if the doctor is culturally sensitive.

In addition to cross-cultural communication issues and the actual contents of the training module, it is important that students are aware of practical issues such as seating arrangements, so that they are prepared when they actually start working. (Bischoff and Loutan's 1999 manual *Due lingue, un colloquio. Guida al colloquio medico bilingue ad uso di adetti alle cure e di interpreti/* ['Two languages, one interview. A guide to bilingual doctor interviews'], although it is meant for healthcare providers, is excellent training material in class with its user-friendly format and diagrams that suggest how doctor, patient and interpreter should sit in relation to each other, and other useful bits of practical advice.) Role play (discussed in chapter 5) is invaluable here, as are videos that show exactly what takes place in an interpreting session.

Language strategies

Translation strategies will be addressed more specifically in the practical sections of this book, but there are a few things we would like to mention here. In the interpreting session the health provider is the information-seeker and thus the 'questioner' in control of the situation. When seeking new information the questioner has a number of questioning strategies at his or her disposal, and these are rarely arbitrary. Doctors and nurses are taught to construct their questions in a particular way, aimed at achieving as much or as precise information as possible, and these formulations should be upheld by the interpreter. Medical training has in recent years encouraged a more discursive, narrative approach to information-seeking, using open-ended questions and encouraging the patients to speak spontaneously for themselves, in the hope that this technique will bring to light contextual variables that may be important diagnostic cues. These questioning formats should be respected by the interpreter (although it is also true that such Q-A formats are in part culturally governed and the interpreter may be a useful source of cultural expertise if and when communication fails due to cultural differences). Because health interpreters are rarely taught the specifics of these language strategies in their preparatory courses, it is always useful to discuss them with the health professionals, when this is possible.

Other issues, common also to legal interpreting, are whether or not a technical register should be lowered for optimal comprehension, if and when to depart from the first-person rule (for example in mental health interpreting or in languages where grammatical forms may lead to confusion, which is possible in deeply hierarchical societies). In highly emotive contexts of fear, anger, offensive language, and so on, describing or reporting may be advisable rather than direct first-person interpretation. In these situations it is easier to convey non-verbal and pragmatic elements that are crucial to the communicative event by paraphrase, rather than by a direct translation where these features cannot be imitated 'theatrically'.

Cross-cultural issues in the health sector

> *A provider gives a non-English-speaking patient a prescription, explaining that it is for some suppositories. The interpreter is too embarrassed to admit that he does not know the equivalent word for 'suppository' in the patient's language, so he uses the word for 'pill' instead. The patient takes the medication orally and ends up in the emergency room.*

In class we try to use vignettes such as this one (preferably humorous ones!) to show the students the possible damaging consequences of poor interpreting in the healthcare setting (taken from the websites: www.xculture.org, www.ncihc.org).

In the health sector we find numerous cross-cultural differences between the host society and the patient's culture which affect the communication process and may seriously hamper understanding, diagnosis and treatment, thus effectively excluding patients from the quality healthcare to which they are entitled.

Cultural differences in interpreter-mediated institutional communication can easily lead to ethical dilemmas, impacting on our translation choices. Examples abound: interpreting a diagnosis to a terminally ill patient – as required in the politics of Western bioethics – with no mitigating communication strategies; views on birth (delivery, diet of mother and baby, breastfeeding, naming rituals); sexuality/ reproduction (fertility, impotence, contraception, abortion); terminal illness (sharing and disclosing bad news); informed consent (the involvement of the family); death (rituals, burial); disease ('taboo' diseases such as leprosy, tuberculosis or in some countries even cancer); suicide; the body (description, touch, uncovering); diet ('hot', 'cold', balancing energies); the description of symptoms (through metaphors, connotations); the expression of pain and the description, perception, articulation of mental health (mental illness, depression, suicide). Last but not least, we have the understanding of confidentiality – with whom should or should not information be shared? (The patient first and foremost? the family as well? the family only?) Differences in attitude towards these issues often hinder communication, collaboration, diagnosis, treatment and patient compliance.

However, the disclosure of terminal illness is perhaps the issue that crops up most frequently. In many countries bad news is not given to the patient directly but through the family, and those Western health-care providers who are culturally aware do often bring the family into the communication process. Many interpreters and patients feel that the way in which this news is broken has a concrete impact on the patient's health, not to mention state of mind. Therefore, many interpreters in Western countries find it difficult to follow their professional accuracy mandate when a doctor – following slavishly the Western bioethical model that demands full disclosure of information to the patient – informs a patient baldly of serious illness. The interpreter thus finds herself in an ethical dilemma, torn between two cultures and professional codes of ethics. Many interpreters do find ways of mitigating the doctor's message without compromising, at least too openly, their own socio-cultural norms and that of the patient, or compromising translation accuracy.

Mental illness is probably the most delicate area of cross-cultural diagnosis and treatment, and may be challenging for a Western psychiatrist in that both diagnosis and treatment – indeed the very understanding of mental illness – varies enormously from country to country, especially for widespread illnesses in the Western world such as depression. The interpreter can have an important function as a cultural informant in these cases, informing the clinician of the role of the family, the perceived impact of bad news on the patient's health, etc. The use and perception of pills and medication also varies enormously across cultures, and doctors should be aware of this in prescribing drugs and making sure the prescription is used correctly (patient compliance).

Verbalizing poor health: euphemisms and metaphors

Many so-called traditional cultures use metaphor far more frequently than Western cultures and give far more weight to its validity as a descriptive tool and as a 'truth-recounting' tool. The description of pain, described through metaphors and images, for example, varies immensely from culture to culture, but so does the social acceptance of 'complaining'. (See Galanti 2000 for a wealth of examples from hospital settings in the US of how the expression of pain is construed culturally.) For the healthcare provider to have sufficient information to ascertain disease by analysing the patient's manifestation of pain, this is not a trivial aspect. Many patients have been over- or under-medicated with pain relief because doctors and nurses have misunderstood how much or how little they were complaining and have misjudged the dosage of pain relief medication.

Interpreting metaphor is not in fact just a question of finding another metaphor or paraphrasing it: the perceived information value, communicative and pragmatic effect or contextual appropriateness of metaphor is highly culture- and context-bound. Symptoms experienced by a patient in a hospital room might well be described through images or metaphor; a patient may be used to describing her symptoms through expressions such as pain being 'hot' or 'cold'. Indeed, the description and self-perception of the body itself is a projection of social, cultural and historical variables.

There is no doubt that socio-cultural factors (such as social and economic class, poverty, level of education) also impact to a lesser or higher degree on provider–patient communication and quality of care, and may exacerbate the already existing communicative obstacle of the lack of a shared language. Effective cultural communicative competence thus improves quality and addresses disparities in healthcare (see Betancourt and Cervantes 2009). There are many examples of cross-cultural differences that significantly affect the perception of health and of illness, its manifestation, its communication (especially to patients and patients' families), its diagnosis, the discourse and language form that surrounds and expresses it, the treatment and cure, the expectations of health in individuals and in the community. These examples are perfect springboards from which to discuss ethical dilemmas in the health sector and can be tied in with more general discussions on cross-cultural communication, on ethics, on translation accuracy and on the professional hierarchy in hospitals.

Ethics

We have discussed ethics at length in chapter 2, so here we will just remind the reader that although there are some fundamental ethical issues that interpreters have in common (accuracy, confidentiality, impartiality, etc.), professional ethics and standards of practice will nevertheless vary from sector to sector, not least because they are governed in part (and this varies a great deal from country to country) by the professional ethics and legal requirements of that particular sector. Indeed, many of the more well-known *codes of ethics* for interpreters relate to specific fields, especially the medical and the legal fields. The professional ethics of each sector – and the complex interplay between the institutional ethics and the interpreters' ethics – will thus vary. In Italy, for example, the institutional ethics would take precedence over interpreting ethics because the interpreting community is still so weak. This means that in the final analysis interpreters will take their orders on how to interpret from the health institution – and also, which tasks they are expected to fulfil. (In those cases where they exist, such as the largest hospitals, this will be via some sort of agreement with an interpreters'/mediators' association).

One of the ethical issues that frequently arises in the health sector is the forming of allegiances between patient and interpreter, although this can be kept in check quite easily by using discreetly distance-creating strategies such as explaining one's mandate and role, limiting physical proximity and eye-contact, maintaining a neutral tone of voice, being generally 'firm' but understanding, not discussing personal matters and avoiding one-to-one conversations. Practising or just discussing these tactics with the students is good attitude training. A more difficult matter is whether to divulge important confidential material if the interpreter has access to information that the patient is not willing to give to the health provider (this may occur if they both come from a small migrant community or if the interpreter has interpreted for the client on previous occasions). One distinguishing criterion here that may help the interpreter make this decision is the degree to which the lack of information will affect the patient (i.e. 'is it life-threatening?'; see Pollard 1997–1998). As mentioned, these decisions rest, in the final analysis, with the interpreter and should be based on her own experience and cultural sensitivity as well as on what she has learned during training.

Preparation

Preparing for a health interpreting session will generally mean revising and possibly updating technical terminology and will clearly depend entirely on the specific contents of the session – which of course the interpreter should be informed of beforehand. Ideally, interpreters will have a technical glossary that they can update using internet or dictionary sources. It is a good idea for students to get used to compiling glossaries while they are still training and to familiarize themselves with field-specific websites and with online translation tools.

A more general preparation will also require being updated on new laws and regulations, on new administrative procedures, and on specifically immigrant-related diseases. These topics should ideally be covered in seminars or workshops provided by the institutions themselves, who have the expertise and the know-how to address these issues in a responsible fashion.

Briefing

The possibility of doctor briefing before the interpreting session and a meeting with the doctor afterwards is very reassuring, because the interpreter herself has the opportunity to clarify misunderstandings and repair any error she feels she might have made. They are also very helpful for the interpreter in that the health provider has the opportunity to discuss details of each case with the interpreter, and to inform her ahead of time of any of any potential terminological, conceptual or psychological difficulties. In this way the interpreter knows what to be prepared for and to pay attention to; she knows, at least to some extent, what kind of reactions she can expect from the patient and how these reactions affect the patient's verbal communication. The interpreter should be informed of any accompanying family members (which may require specific interpersonal social strategies), the patient's age, sex and knowledge of the host language. If the patient speaks a language that includes different dialects, this is also important information. The interpreter can then also inform the doctor about her practical issues such as if and how to interrupt, where to sit and whether to use the first or third person address mode. And of course the interpreter can signal any potential cultural misunderstandings, interpersonal dynamics between the three parties or between any other parties the patient might bring along, to inform the doctor or nurse of the patient's signs of unease, discomfort or pain, which may be lost upon the service provider.

Students should also be made aware that briefing is not always realistic in a hospital or clinic where time-pressure can make life very difficult for all parties involved. Nevertheless, briefing should be mentioned and if we are able to motivate our students to insist on briefing sessions when they actually do start working as interpreters, this may be an indirect but effective way of educating future service providers.

3.2 Interpreting for legal services

3.2.1 Organization and structure

We have already mentioned the importance of interpreting in the legal sector for the effective implementation of justice in any country that hosts migrant communities, tourists, foreign students and other non-host-language speakers. Indeed, in those countries where the organization of language services is poor, it effectively jeopardizes the rights of the individual actors, the effective running of the judicial system, but also the safety of the local community. It is good pedagogical strategy to inform students of the legal services structure in their own country, of the potential risks they and their fellow citizens may run in their local communities if there is no provision of crucial interpreting services for the police and the courts. This then, is the global social and institutional aim of multilingual communication in the legal sector.

Arguably, even more so than the health system, however, the organization of a national legal system is highly culture-specific. We will therefore not use this forum to discuss these specifics but urge trainers to discuss with their students the organization of legal institutions in their own countries and if, where and how interpreters are positioned in this system. This should then lead to a discussion of the various phases of the judicial process (investigation, arrest, custody, trial, sentencing, prison life, etc.) that require different communicative techniques from the various actors (undercover agents, police officers, police questioners, lawyers, magistrates, prosecution, defence, judges), and the communicative format required in each of these phases. Knowledge of the historical developments and regional variations of legal institutions (clearly, quite an ambitious programme) should be adapted, realistically to the level of the students and the objectives of the course.

Communication processes such as questioning witnesses or suspects at the police station, cross-examination in court, wire-tapping, and so on, will require very different techniques from both primary actors and the interpreters involved. These techniques may vary a great deal for the different languages and cultures involved. Furthermore, in some countries the *role* of the interpreter, not just the communicative strategies, may vary greatly from phase to phase – they may be far more involved and proactive in the preliminary investigative phase (acting as cultural informants and advisers) than during the trial phase.

Like 'medical interpreting', 'interpreting in the legal sector' is also an umbrella term for interpreting in a number of different institutions that also require different strategies and that cover a range of registers: police interviews, lawyer–client interviews, detainment centres, juvenile courts, the preliminary investigation phases, courtroom proceedings and cross-examination, various tribunals, the Ministry of Justice, and so on (see Mikkelson 2000; Hale 2007; Gentile, Ozolins and Vasilakakos 1996; Berk-Seligson 1990). Each of these settings will require, then, different terminologies, registers, interpersonal skills and discourse and translation strategies.

The participants in the legal system also vary from system to system and country to country, not only the magistrates, lawyers, the jury, the prosecution and defence teams, but also the court clerks, administrative personnel and technicians. Again, trainers should present the various actors in their own system and ideally provide a comparison to systems in other countries, in the particular L2 if one is working with a specific language pair. (This is good for translation purposes and helps create clarity where there are potential pitfalls such as for 'false friends'/'faux amis' and unfamiliar terminology). Translation can be an enormous problem here because very often the legal systems and the legal actors do not correspond and one may need to create neologisms and/or paraphrase accurately. (The legal dialogues in chapter 6 give plenty of opportunity for terminology practice.)

We found it useful to give visual support (diagrams) of the layout of a court hearing or police interview or a courtroom, the various actors, the seating arrangements, and so on. There are many sources on the web that have good visual back-up, and others that give the bilingual or multilingual terminology of the courtroom and police rooms. We also like to give a brief indication of the chronology of a case (from arrest to sentencing). If time does not permit a description of these, then leaving handouts for the students is sufficient (with this terminology and conceptual framework they can write their own dialogues and conduct research independently).

Communication structure and strategies

Social and emotional context
In the legal setting too, what we have called the 'emotional context' is important. Patients are by definition vulnerable, because their reason

for seeking the services of health institutions is reduced, or perceived as reduced, health. In legal settings too the actors are often – but not always – in a vulnerable situation and/or frame of mind. What exacerbates the vulnerability, however, is that the atmosphere of the police station or courtroom is very different from that of the hospital or health centre, and the relationship between the actors thus changes radically. This is due to the authority invested in them by national law, but also the expectations and traditions of the local community. Police questioning may also evoke memories of aggressive interrogation techniques, especially for applicants for political asylum, for whom it may bring back fear of torture.

Interpersonal rapport and power asymmetries

In addition to the visual layout of the various settings, becoming familiar with the actors, and so on, it is also useful to give students an idea of the interpersonal dynamics between the participants and to remind them that it is the police officer/ judge who is in control of the situation and must ultimately take responsibility for what takes place. The tenor of the rapport with police officers and judges is important to the client. And the fact that this rapport is necessarily mediated through the interpreter is a potentially daunting responsibility. The interpreter's rapport with the client is also important and must be kept within the bounds of the prevailing code of ethics (if and when this exists and/or has been communicated); for example, the interpreter must not enter into conversation with the client any more than is necessary for required communication.

During the preliminary phases that are so crucial in gathering evidence and establishing the credibility of the suspect, interpersonal strategies are equally important and the interpreter should be aware of the primary actors' overall aims, which may not necessarily be obvious in that the questioning techniques may be specifically geared to gathering information in a way that is not immediately clear to the outside parties (trick questions). During this phase, factual information is established and used as a foundation during the trial to arrive at a verdict. The questioners will base their impression of a person's credibility on language and para-verbal features such as presentation of content, coherence, delivery, manner, register and style. The interactional rapport in witness interviews will be different again, and the questioner's objective more transparent – less aggressive, manipulative and confrontational. The rapport is more that of a 'client' and the witness is likely to be less defensive although often very fearful, a state of mind which may also hamper straightforward communication.

The questioning techniques vary from country to country, and if at all possible, it would be useful for the trainer to find out, from interviews with suspects, what these are in the student's own country and local community. Questioning techniques and linguistic and non-verbal communication are again highly culture-bound and are also deeply affected by the level of hierarchy in the tradition and custom of that particular community. Question-answer sequences will be deeply governed by the asymmetries of power, not least who is seen to be in control of the communicative act and the degree of authority vested in him or her by the culture and by the national authorities. These interpersonal relations between the various actors can then be practised in role plays with the students, taking care to use the correct and appropriate forms of address and politeness.

Strategies

We have mentioned that each of the phases in the legal process – from investigation to sentencing to prison life – are expressed through, and will require, different communicative strategies. It is perhaps the question and answer sequence of information gathering and ascertaining/validation mentioned in the previous paragraph that most marks legal communication in all its phases. This will be more spontaneous and unplanned in the early investigative phases, but become almost ritualistic and rhetorical, planned and orderly in the final phases in the courtroom. (Hale discusses some of these in Hale 2007: ch. 3).

Commissioner D. Rombouts from the Antwerp police gave an excellent talk at an international conference organized at Lessius University in November 2009 on legal interpreting and translation, on whether or not interpreters need to know police interviewing techniques (Rombouts 2009). Question format seems to be an important element in the gathering of information. Examples include:

- Open-ended questions
- Closed questions
- Multiple choice questions
- Leading questions
- Probing questions
- Opinion/statement questions

- Bait questions
- Behaviour questions.
  (From Rombouts 2009)

The 'good cop, bad cop' police questioning routine is one which is familiar to us from films. The primary goal of interviews is to get the witness to *volunteer* information rather than extract information, Rombouts says. There are various ways in which to facilitate this – for example by allowing the speaker time to offer a 'free story', to avoid interrupting and to allow for long pauses, as well as other more focussed techniques, including lie-detection techniques. Rombouts lists possible pitfalls for the interpreter: lack of an introduction to clarify roles at the start of the interview, failure to allow for pauses, failure to translate everything, inaccurate translations of lexis or syntax, and – importantly – the failure to follow the system of open-ended questions. Practising these question formats with the students – ideally working them into the dialogues – would be extremely beneficial to prepare them for work as legal interpreters.

Lastly, we ask the question whether para-verbal elements such as body language, tone of voice and gestures should be reproduced, and if so – how? (see Mikkelson 2000: 50). This is a good discussion point with the students and both useful and motivating to practise in role play in class, suggesting and co-constructing creative solutions together.

Clarification and simplification

One question trainers are often asked by students is whether or not the interpreter should simplify difficult language to the client/defendant. One might throw the question back to them and ask what happens in a parallel monolingual situation? Does the officer or judge himself paraphrase his own words? This is arguably one of the more delicate (rather than terminologically difficult) aspects of the legal interpreter's role – to evaluate whether or not it is her task to adapt the service provider's language to the client when at the same time she is expected to interpret accurately and fully everything the speaker says.

Although it is not an interpreting mode as such, we would also like to mention here *sight translation*, which may be required of interpreters in legal settings. An interpreter may be asked to sight translate documents, and sometimes to translate/write written statements that are kept for the record or may be used for exhibits in court. Here, terminology and language register become of crucial importance. Because of differences in the legal institutions from country to country, the interpreter must often paraphrase or create parallels where there are no ready correspondences. Introducing simple sight translation tasks in class is therefore an excellent preparation. (See chapter 5 for ideas on how to teach sight translation.)

Repetition and repair strategies

In all settings, the students should be taught to ask for repetition if the utterance(s) are not clear, and they should practise doing this discreetly. However, it is much more difficult to interrupt actors in this context than in a doctor's office, and the students need to learn slightly more assertive (firm but polite) turn-taking techniques. Self-confidence and an awareness of their own crucial role and responsibility in this situation are important. When the interpreter is aware of having made a mistake, employing *repair strategies* is equally crucial. In court this is particularly challenging due to the innate professional hierarchy, but Edwards suggests that corrections can be made by waiting for a break and approaching counsel to ask if the question might be asked again or alternatively, by asking to speak to the judge after the session (Edwards 1995).

How the interpreting services are organized

As in the medical sector, the recruitment of interpreters is extremely culture-specific and each trainer will have to describe the practice in her specific local situation, specific to her country, city, town and institution. In those countries where untrained interpreters are recruited on the basis of 'knowing the language', the dangers of using ad hoc, unskilled interpreters should be discussed at length. This may even lead to stimulating and constructive discussions on how to improve the prevailing conditions, making the students feel that they are active members of the community, contributing to the improvement of services and conditions for themselves and their fellow citizens.

3.2.2   Sector-specific contents

Terminology

Although ideally this should not happen, the level of terminological expertise needed does actually vary from country to country, because in those countries with poorly organized interpreting services and low expectations as regards interpreter qualification and performance, legal actors will not always expect interpreters to translate everything that is said in the interviews or during the hearing, but only the questions which are directly addressed to the foreign language speaker (and the replies), ruling out the longer technical passages that contain extremely complex texts. Although this might initially be reassuring for students in those countries, it does have a number of ethical consequences, however, and these should be spelt out: when only excerpts from the trial are conveyed, depending on the judge's decision to ask for a full translation or not, one might say that the defendant is not fully cognizant of everything that is going on at his own trial and therefore is disempowered. The defendant/witness must place enormous trust in the defence lawyer, interpreter and judge and one might question whether or not this trust is always deserved.

Annotated field-specific glossaries can be assigned as term papers, and this is a good way of encouraging students to do terminological research. Authoritative legal dictionaries are obviously the most reliable sources for understanding technical terms and for translation, but there is a wealth of material available through the internet, often through local police or legal sites (see, for example, the Australian website for *The Interpreter Handbook* (http://www.mrtrrt.gov.au/Tribunal/Interpreter Handbook/17_Guidelines_for_interpreters.asp) that discuss various interpreter-related issues, and many explanatory videos on legal-related matters (for example, in the US: http://courts.michigan.gov/scao/services/access/InterIntro.htm, on the arraignment, bail, the hearing, sentencing and the trial).

Cross-cultural aspects

We have mentioned the question-answer format of police interviewing in many Western countries today. It is important to remember, however, that this may become a communicative ordeal for people who are used to other questioning formats, and their answers (both in formulation and content) may not be delivering what the questioner expects. This communicative misalignment may even be mis-read by the interlocutor as untrustworthiness. For example, a witness might use roundabout expressions or paraphrase which might be perceived by the judge as 'inaccurate'. Or else, metaphor might be used to express factual information, which again might affect the judge's or jury's understanding of the truth value of the utterance, and therefore the judicial outcome. Similarly, silence and hesitancy (not just pauses in the question-answer mode) can be interpreted as shiftiness or non-cooperation, while the person might just be thinking of an answer that will best please the interlocutor. Aiming to please and acquiescence, especially towards an interlocutor who is in a hierarchically superior position either socially or institutionally, is an important cultural trait that is not shared by most Western societies. On the contrary, it is regarded with suspicion, even deception.

Another crucial cross-cultural pragmatic communication feature that appears frequently in the legal setting is the use of kinship terms as identification parameters (see Garzone and Rudvin 2003). Time and time again the use of 'brother', 'sister', 'cousin' or even 'mother', 'father', 'grandfather' is translated directly into the major European languages and no account is taken of the semantic and pragmatic second-level meanings of these terms in other languages. In Pakistan, 'brother' can also mean 'cousin' (distinguished respectively as 'proper brother' and 'distant brother'), indicating the strength of the extended family system and solidarity between kin (see Rahman 1999). The use of the father's or husband's name as the surname in many Muslim cultures can also cause unnecessary misunderstandings, suspicion and confusion when a witness, defendant or patient is asked to identify himself or his kinsman – or indeed to identify another person in terms of their kinship relations. Kinship terms are also used for respect (e.g. 'older sister' for a woman one's own age or slightly older and the ubiquitous 'auntie' or 'uncle' in both Hindi/Urdu and English, especially but not exclusively used by children, and not necessarily for a relative). We see that the potential for misunderstanding is enormous and potentially very serious. We feel therefore that there are times when the interpreter should indeed – in an appropriate manner ideally decided ahead of time with the various actors – act as a cultural informant. This too can be practised in role plays.

We also feel that it is useful for the students to be able to empathize with the non-host-language speakers, in order to better understand their comprehension difficulties and provide a more accurate rendition. However, this empathy should not turn into excessive or potentially intrusive bonding. One light-hearted and humorous way in which we have sometimes tried to communicate to students the confusion, fear and vulnerability of a newly-arrived migrant who finds himself, perhaps even without knowing why, in the highly threatening environment of a courtroom without understanding either the culture, language or institutional objectives, is that of quoting from the trial scene in Lewis Carroll's *Alice in Wonderland*, which so many people educated in the West or in Western school systems are familiar with from childhood. Carroll's satire of the utterly nonsensical aspects of Victorian society is brilliantly portrayed through the chaos and anarchy of the courtroom scene, the outrageous and arbitrary nature of the accusations, Alice's confusion and sense of total alienation and fear. This fear, bewilderment and confusion can be used as a metaphor for the state of mind of a twenty-first century migrant to the West in a language, country, system and institution that is totally alien to him or her.

Ethics

We have already discussed the issue of translation accuracy, which is no doubt the most important ethical imperative in legal interpreting. We have also talked earlier about confidentiality, but would like to stress here how important it is for the effective running of the judicial process, for the safety of the individual actors, and for the interpreter's own safety, to maintain confidentiality. However, it may also be useful to remind the students how tempting it can be to discuss interesting cases with colleagues, friends and family; indeed, it is not only interesting but may act as an emotional safety valve in particularly violent or emotional cases. It is reassuring also to remember that we share this frustration with most of the other professionals working in this field – lawyers, police officers, judges. There are ways in which interpreters who work in particularly psychologically and emotionally stressful contexts can share their experiences with colleagues by refraining from mentioning names and details that can lead to identification of the people involved. These are useful strategies by which interpreters can uphold the code of ethics without feeling submerged by guilt, frustration or vicarious trauma.

Preparation

Again, preparing for a session at the police station, the lawyer's office or courtroom requires more than anything else terminological revision. The interpreter should ask for as much information as possible, to be able to prepare for the assignment, but this too will vary greatly depending on the time available, the nature of the case, the institution and the country. The students should be encouraged, as future interpreters, to be assertive on this issue. It is always a good idea to familiarize oneself with the names of the people involved – both suspects/witnesses and legal actors – because they may crop up repeatedly, cause pronunciation difficulties and are often difficult to remember (practising memorization with lists of foreign names is a very useful classroom exercise). The acoustics of the courtroom may be such that any potentially difficult lexical items that cannot be inferred from the co-text and that may cause pronunciation difficulties, are more easily understood and reproduced if they are already familiar. An interpreter is also entitled to know at least the generalities of the case and the charge, the dialect, age and sex of the foreign language interlocutor, and also if they speak the host language at all. It may be useful to remind students that culture- and institution-specific items are often more effectively communicated if left in the original, for example the names of offices, work and residence permits, national laws, and so on.

Briefing

In our experience, only very rarely do interpreters have the opportunity to speak with their interlocutors beforehand, and if so it is only a very cursory word with the lawyer, magistrate, clerk, etc. to verify details, or to ask questions about the practical issues involved. If possible, the interpreter should speak to the suspect/witness for a few minutes to establish that they actually speak the same dialect/language (sometimes false passports may create misunderstandings about citizenship and the language spoken). If and when serious language-based or cultural misunderstandings have arisen during the session, the interpreter should insist on a post-session briefing with the person responsible for the session and try to explain as fully as possible her views on the matter.

3.3   Interpreting for the business sector

3.3.1   Organization and structure

The organization of interpreting services in the commercial sector is very different from that of public institutions precisely because 'business' is neither a public nor a monolithic institution, nor is it rigidly structured. Indeed, the very nature of commerce is based on flexibility and private enterprise. Due to this flexibility and individual decision-making, it is the individual manager or the individual company rather than the town, state or region who contacts the interpreter for each specific assignment. The company may have its own interpreter or may use a linguistically competent secretary or other employee. Depending on the agreement between the companies, on where the meeting is taking place, on the companies' budgets and on their awareness of linguistic or cultural issues, the interpreter may be commissioned by either of the parties (often the company operating in the host country), or the companies may have an interpreter each. When the interpreter is paid by one of the parties and that party acts as if the interpreter should have primary allegiance to them, the interpreter may feel compelled to act on that company's behalf. The interpreter may be asked to provide opinions about the other party (e.g. sincerity of their offer, likelihood that they will continue to negotiate) which may seriously jeopardize the interpreter's perception of, and adoption of, the principle of impartiality.

Participants: variety of role and settings; multi-tasking

A peculiarity of business interpreting, more than interpreting for the public sector, is that there is a higher degree of multi-tasking, and there are times when interpreters are expected to fulfil numerous tasks that have nothing to do with interpreting or translating. Furthermore, there is a far wider range of settings, from the office to the conference hall, factory floor, aircraft, restaurant, construction site, etc. as well as a variety of activities. There is also a far wider range of formats and registers such as:

- formal meetings which would encompass introductions, welcoming speeches, protocol
- smaller meetings of groups or one-to-one meetings at various levels of formality
- discussions of main points in contracts or work plans
- technical group meetings
- visits to relevant sites or institutions
- extra-curricular activities such as tourism or recreation, meals and formal banquets, after-dinner speeches.

It is easy to see then, that the interpreter's role in the business setting is is varied. This applies not least when the interpreter is travelling abroad with a delegation and attending meetings or visiting trade-fairs (the interpreter, we must remember, is often the only person who knows the language). It also applies when the interpreter is acting as public relations officer or secretary and collecting documents, writing minutes, making hotel reservations, paying bills. The variety of settings will also require the interpreter to switch between consecutive and chuchotage modes.

Switching from one task/role to another is tiring, requiring constant concentration – she is 'on-duty' round the clock. Often, the work continues even in the evenings, if she has to translate after-dinner speeches; another source of stress is competing demands from delegation members.

Although new recruits to interpreting tend to work in many areas to gain experience and earn a living, interpreters who work primarily in the business sector have very different backgrounds from those who are trained to work primarily with migrants in public institutions. In many countries, interpreters who work in the business community will often have a general language background from modern language faculties. Their language combination will often be languages such as English, French, German, Russian, Spanish – the major European languages and the languages that tend to be culturally and commercially predominant. Even with non-Western interlocutors from Japan, China or Arabic-speaking countries, English is frequently used as a vehicular language. In public institutions and community-based services involving migrants, however, the most commonly used languages come from a wide range of non-Western and Eastern European languages, languages that represent current migration trends, very different from the pool of languages used in the commercially driven private or public sectors. The interface between these language combinations is the major lingua francas, especially English. Thus, interpreters whose language repertoire includes one of these lingua francas can use them in a wide variety of applications.

Communication structure and strategies

The lack of institutional asymmetry is one of the features that most clearly marks the difference between the commercial sector and the public sector and which most affects communication. The participants are not legally authorized representatives of public institutions but private individuals meeting in a professional context for a specific objective. We have indeed suggested that in all settings, communication form and discourse strategies are heavily influenced by the primary purpose of the encounter.

Roughly speaking, negotiating could be defined as reaching an agreement between parties who begin from different bargaining positions. Unlike the public sector, in the business setting this will generally be *negotiation* of some form (and therefore very often will include potential, covert conflict, although this is often masked by politeness and camaraderie in the aim of reaching a negotiated result that is favourable to oneself). Business interactions are in fact distinguished from other interactions described in this book by the fact that reaching an agreement through negotiation is their central objective (so-called transactional conversation). And the very structure of the encounter will be influenced by interpersonal negotiating strategies varying from positive face (friendliness, flattery, etc.) to open hostility. Thus the range of registers and terminology will vary greatly (as in other settings mentioned here) from introductory pleasantries, greetings, small-talk to highly technical contractual or other terminology. This range of registers requires numerous skills apart from terminological and technical competence – skills such as diplomacy, tact and politeness, but also assertiveness, directness and of course an acute awareness of the cultural codes at play. We have tried to reflect this register variation in the dialogues we provide in Chapter 6.

Teachers' and students' experience as business interpreters at trade-fairs

Many students start their interpreting careers at trade-fairs, often working for free, multi-tasking and working long hours. The context of trade-fairs varies enormously from cosmetics, to motorbikes, to the building sector, fitness machines, ice-cream, and so on, and this is demanding but also stimulating and enriching for the students. Referring to one's own working experience to provide students with 'hands-on' and real-life settings is useful here.

Role play and dialogues

It is also useful to provide students with 'real-life' examples of transcribed conversations that demonstrate cross-cultural misunderstandings. Although we have not used it for our dialogues, much inspiration can be drawn from Craig Storti's simple and user-friendly (1994) book *Cross-Cultural Dialogues: 74 Brief Encounters with Cultural Difference.* Storti provides a short analysis of each dialogue and shows us why it 'went wrong'. Although the book is not recent, these dialogues are an excellent way to illustrate cross-cultural differences expressed in natural conversation. They can also be used to illustrate the cultural models of Hofstede and of Trompenaars and Hampden-Turner mentioned in chapter 5. There are many websites that provide insights into intercultural training courses for businesses, many of which however are accessible only on payment. We have found two that are particularly sophisticated: www.culturewise.net (CultureWise) and www.countrynavigator. com (Country Navigator).

3.3.2 Sector-specific contents

Terminology

We have suggested above that business discourse is marked by a wide variety of fields and registers ranging from the very informal to the formally and deliberately polite, to the contractual and potentially conflictual, to the highly technical. The element that is missing compared to the public sector is the institutional-bureaucratic language. Of course, during the technical phase of the encounter the terminology may be very complicated indeed. However, although technical jargon is often feared as the most difficult part of the translation process and is challenging for an interpreter, the primary interlocutors often feel most at their ease when using this terminology rather than 'everyday' terminology, precisely because it is their area of expertise. Students, as future interpreters, are often reassured by the fact that interpreters can ask specialists to clarify or explain simply, spell out a word, write it down, draw a diagram, write the formula, etc. The other interlocutors will probably understand a technical term if it is spelt out or presented in the form of a diagram.

Cross-cultural aspects

Cross-cultural issues are also crucial in international business communication, as the vast literature on intercultural business management demonstrates. These cross-cultural issues are different from the medical and legal settings in some fundamental ways though: the 'typical' actors are not migrant and doctor, lawyer, teacher, social service worker from the host country, but businessmen and women from two or more countries often using a vehicular lingua franca (see Garzone and Rudvin 2003 for examples). Secondly, the 'power alignment' which affects interpersonal relations and discourse strategies changes – it tends to be a conversation 'between equals' on the same socio-professional level, both being experts in the field, rather than the layman versus expert rapport typically found in public services. Thirdly, the objective of the communication session is no longer a service guaranteed by the state, but a private negotiation whose outcome is more unpredictable. In other words, the outcome of the negotiation is not something the clients are constitutionally entitled to, but one that must be achieved, and one that depends on the performance of all the actors, including the interpreter. The interpreter's mandate and alignment changes as a result of this and may affect her translation strategies and code of ethics, not least in terms of who she is accountable to (generally who she is paid by).

There is an enormous amount of literature that addresses cross-cultural differences in the workplace and in business but we will here limit our suggestions to how these differences affect the corporate identity of the participants and the resulting interpreter-mediated communication. We have mentioned here some of the cultural features that we have found to be most helpful in explaining how misunderstanding can come about, and by the same token how they can be pre-empted or mitigated by the presence of an interpreter as an expert cultural and linguistic informant.

Corporate versus ethnic culture: management styles

We like to remind the students that the workplace, like a public institution, can be considered a micro-culture ('corporate culture' in a very wide sense of the word) unto itself, as yet another cultural sphere that must be negotiated by the various players in the game. In terms of cross-cultural issues, it is clear that management styles are highly culture-bound. Not only are they culture-bound in an ethnic sense, but each company has a particular 'corporate culture'.

The organizational structure and conduct of a company will obviously affect management communication styles between staff, and between that company and other companies, for example a large multinational and a small family-run business. The interpreter must have good knowledge of the negotiating styles of the different parties (both national and corporate styles) and recognize communicative features such as the degree of directness, willingness or ability to reach a decision or make that decision known during the meeting. This knowledge must include the use of agreement or disagreement, the role of ceremony and protocol in negotiation, the expectations of sequencing of elements in the negotiation, and expectations relating to dress, food, body language, age, gender, hierarchy, address, titles. The interpreter will have to accommodate each of these styles and the dynamics that arise as a result of the interface between different management styles, e.g. whether obliging, dominating, conciliatory or aggressive. When emotions run high (frustration or anger are often a result of the nature of the negotiating process), the interpreter must make a number of difficult decisions: do you tone the language down, modify the emotivity to facilitate communication? (Straightforward interpreting of anger or rudeness may kill a negotiation rather than allow it to continue.) It is also interesting to note that the interpreter is often drawn into the communication between the parties when tension arises, as it so often does when different management styles are adopted (see Garzone and Rudvin 2003 for an example of this).

Management styles can indeed vary a great deal; Zhuang (2009) shows how management styles can be integrating, obliging, dominating, avoiding or compromising. His schema of management styles in different cultures is easy to use in class; the example we are using here is that of Chinese versus American:

- Avoid confrontation versus confront the issue
- Implicit versus explicit
- Face concern versus focusing on the issue at hand
- Mediation and resorting to a third party versus relying on one's own efforts
- Compromise versus 'win or lose'.

Although this representation is no doubt highly schematic, it is easy to see how even very general patterns of cultural difference can create significant tension in business negotiations. We have found that working these cultural schemas into dialogues for role play, and encouraging the students themselves to invent dialogues on the basis of these differences, can be very stimulating.

Ethics

The issue of ethics has already been addressed above: when the interpreter is hired and paid by one of the parties, this may create serious difficulties for the interpreter if that party acts as if she should have primary allegiance to the hirer.

Preparation

Just like the other sectors, preparation for an interpreting session is mainly focussed on field-specific terminology, which in this sector may be very technical indeed. A good glossary, either a self-compiled one or an online technical glossary, is essential. As in the other sectors, getting as much information as possible from the interlocutors before the session is an enormous help in providing as much contextual, technical and terminological information as possible. Collecting brochures and other technical material from trade-fairs, previous encounters, or from the web is also a great aid.

Briefing

Perhaps the most useful information for the interlocutors that the interpreter can provide during a pre-session, or alternatively, post-session, briefing is cultural information that will help the meeting go smoothly and help avoid any potential conflict, misunderstandings or *faux pas* that may cause offence to the other party and jeopardize the outcome of the immediate negotiations or the long-term relationship between the parties.

3.4 Interpreting modes: consecutive, chuchotage, dialogue versus working in booths

In chapter 1 we looked at how the activity of interpreting can be classified by virtue of its area of application and/or by virtue of mode. We very briefly touched upon various interpreting modes, which we will now develop in the last section of this chapter so that they can be taught alongside the contents and structure of the three sectors we have examined here. Many of the skills required are common to all modes of interpreting: note-taking, memory, terminology, fluency and speaking ability, listening ability, concentration, etc. How much both students and trainers will know about interpreting techniques will, however, vary a great deal, so we will try here to give a very basic introduction to the various formats used in conference interpreting and in the forms of interpreting we are dealing with in this book, for those readers who may be unfamiliar with these notions.

Conference interpreting: simultaneous and consecutive

In international organizations, conference interpreters, at least in theory, employ the *monologic* mode, namely translating only from language x into language y (the rule being from the second language into the mother tongue) for simultaneous interpreting. For a number of practical reasons, however, conference interpreters in many countries may work from and into their first, second and possibly third languages. Simultaneous interpreting takes place in a booth that contains the necessary equipment to translate through a microphone and head-sets almost in real time, with only a very short lag between the original speaker's utterance and the interpreter's rendition. Simultaneous interpreters will (if possible) work in teams so that they can take a break when concentration begins to wane. Teams are made up of two or three interpreters, depending on the venue (private market or international organization). They will often have support material with them in the booth – dictionaries or specialized glossaries in book or electronic form (laptops with internet connection) and they can consult each other if necessary. Often they will have the speaker's paper beforehand, which gives them time to prepare the terminology – but of course the speaker may deviate from the script at any time. The setting will generally be academic, scientific, technological, political etc., and the meeting is usually held in a confined, well-equipped space with an audience which is familiar with the subject matter and the specific terminology. Languages are generally restricted to those which would provide a natural setting for a conference (usually not languages of limited diffusion, nor when a lingua franca is widely spoken and which therefore could be used in place of interpreting)

The other mode employed in this setting is consecutive interpreting. In this form of interpreting the speaker's utterance is limited to 'chunks' of speech that are then translated by the interpreter. The chunks of speech need to be short enough to be manageable and recalled by the speaker in order to give an accurate rendition, but well-trained interpreters can handle quite long chunks of speech. The exchange between speaker and interpreter will require a number of discourse cues signalling the end of a chunk of speech or interrupting the speaker when the length is excessive. The consecutive conference interpreter is trained to take notes to aid recall, especially for dates, numbers, lists, and so on.

Dialogue or face-to-face interpreting

In dialogue mode the interpreter will use a simplified consecutive mode in both directions, that is, not only from language x to language y, but back again to language x, switching back and forth from and into both languages at short intervals. Indeed, this is one of the peculiarities of face-to-face, or dialogue, interpreting. Typical of this simplified form of consecutive interpreting is that its duration is normally shorter (in that it is a conversation and not a lecture) and somewhat slower, as it follows a conversation with naturally occurring spontaneous speech rather than a prepared talk. It is easier for the interpreter to interact with the speaker in the dialogue format, to tell him to stop or paraphrase what he has just said, which of course is not always possible in the much more formal conference setting (or in court). Moreover, the discourse and interpersonal strategies are very different: the speech form used in a patient–doctor consultation, with a wide variety of registers from chatty to technical, in a small room containing perhaps only three people, is very different indeed from translating a prepared talk on engineering in orderly chunks to an attentive audience listening carefully and completely familiar with the subject.

Making the distinction between simultaneous conference and dialogue interpreting is useful in that it allows us to present and discuss the communicative and speech-related aspects of interpreting bi-directionally versus mono-directionally and to a large audience with a prepared speech versus (semi-) spontaneous talk with a small group of participants. This usually leads to a cursory mention of elementary notions of conversation analysis, proxemics, the 'speech participant' terminology of Goffman (1981), as well as more specifically institution-related aspects (professional registers, mixed registers, interpersonal relations and hierarchies).

Note-taking

In consecutive conference interpreting courses students are required to build up their own personal note-taking systems, and start from analysing the source speech, so that their notes become a visual representation of their analysis. This means identifying the communicative function of different parts of the speech in order to recognize the main ideas and the secondary ones. Consecutive interpreting students are often told by their teachers: 'Note the ideas and not the words!' Ideas can be defined as 'parts of the message' (Thierry 1981), which tell us 'who did what to whom/what' or the underlying meaning of words or expressions. Consecutive interpreting is in fact about identifying the core message amongst all the other information, that is, recognizing the backbone of the speech, taking adequate notes, and then reproducing the full speech in the target language. The note-taking structure should give the students the right prompts. It is not always possible to include note-taking in dialogue interpreting classes to the same extent that it is taught in conference interpreting, but a brief introduction to note-taking and guiding the students in their note-taking during exercises is always useful. We have found Andrew Gillies' book (2005) useful and user-friendly in the classroom.

Dialogue interpreting adopts a 'short' or 'simplified form' of consecutive interpreting in brief chunks that follow the conversation. In chapter 5 we repeatedly insist on the importance of training the students to interrupt the speaker, either to bring the chunk to an end in order to remember and reproduce accurately what has been said, or to ask questions to clarify anything that might lead to confusion or inaccuracy. If the student decides to ask a question, she has to make sure it is in the speaker's language (it is actually very easy to get confused....), brief and to the point. However, some note-taking symbols can be useful to clarify ambiguities. A big X or another letter and/or symbol to signify a word that needs clarification (Gillies 2005) or a question mark next to a key-word may suffice as a mnemonic aid.

Chuchotage

A peculiar form of simultaneous interpreting is also used in community interpreting and in business interpreting, namely whispered interpreting (also known by the French name *chuchotage*) which is either done by whispering or by speaking in a low voice (Pöchhaker 2003). This form of interpreting implies simultaneous, real-time interpreting. The interpreter processes the speaker's utterance as she whispers the rendition into the receiver's ear with a minimal time-lag. (The difficulties are, understandably, enormous, both the acoustic problems and the neuro-linguistic language processing.) The interpreter may subsequently use a normal consecutive form to translate back to the original speaker in a 'mixed mode' format. This form of interpreting cuts down dramatically on the time used, compared to a straight two-way consecutive interpreting format, but is very challenging for the interpreter. In some countries this format is used widely in the court and in business settings. Chuchotage is a particularly difficult interpreting mode, and we suggest that students experiment with it, if time permits, towards the end of the course, when they already have interpreting experience.

Each sector uses a suitable mode

Although they all employ a form of dialogue interpreting, each of the sectors above tends to use a mode that is suited to its specific requirements. In the hospital, as in the police station, a more relaxed form of simple consecutive is used, whilst in business settings it will often vary between consecutive and chuchotage in a 'mixed mode' format. Courts also often use a mixed mode format which cuts down dramatically on the time used, compared to a straight two-way consecutive interpreting format.

In the *health* sector, simple consecutive is the default interpreting mode, as it is in the other sectors in this chapter. Interpreters in the health sector – like *business* interpreters – can and should adjust their interpreting strategies to the specific requirements of the situation, and this may require switching to chuchotage or copious note-taking, or frequent interruptions, etc. The default use of the first person singular may also be abandoned if there is the risk of misunderstanding between the parties, especially if briefing – where the interpreter can explain her strategies – is not an option. This may happen in mental health interpreting, for example. It is useful to remind the students that it is important to adhere to the general interpreting rules and codes of conduct, but that the overarching objective in this sector is the well-being of the patient (achieved through diagnosis, treatment and patient compliance) and that in the final analysis it is this, and not exclusively our own professional standards, that should be foremost on our list of priorities.

As mentioned, the default form of *legal* interpreting is the simple consecutive, sometimes combined with chuchotage in the courtroom (towards the foreign language speaker). Some courtrooms are also equipped with electronic equipment that allows simultaneous interpreting, and some police stations are now experimenting with telephone interpreting. Interpreting strategies should thus be adjusted – ideally in collaboration with the primary parties – to acknowledge the specific contextual requirements. Mention should also be made of the use of telephone interpreting in both the health sector (especially the emergency wards) and the business sector (there is very widespread use of normal two-way telephone conversations through an interpreter or a secretary with the needed language skills), although this mode is difficult to practise in the classroom.

4
Teaching Methods and Objectives: Course Structure

4.1   You *can* teach an old dog new tricks: adapting old methods to new challenges

When we started teaching together in 2001, community and public service interpreting as a discipline at university level was still practically unheard of in Italy. 'Language and cultural mediation' was already an established *profession* in healthcare and in educational and social institutions, but not at that time a university discipline. Ad hoc solutions were being adopted in these sectors and untrained mediators used widely, but at least there was an awareness of the need for training and of an existing professional community. In the business and legal sectors, where untrained interpreters were – and still are – recruited largely on the basis of 'knowing the language', there was much less awareness of interpreters/ mediators as a professional community and of the need for basic training. Our first priority as we started this course was therefore to *widen* the parameters to include more settings, fully cognizant of the fact that a semester or an annual module of 60 or 80 hours spanning three or four different settings is not nearly sufficient to train a qualified interpreter.

This decision was also based on recognition of the rapidly growing number of migrants that was leading to an increase in the need for language services. The widespread notion that the solution to this 'temporary' phenomenon was the expectation that immigrants would learn the host language sufficiently and sufficiently quickly to function adequately in the host society seemed to us unfounded then, and seems even more unfounded today with a continuing increase in immigration and dramatic peaks coinciding with various events in the global political scenario, not least in nearby North Africa.

We have mentioned above that the need for interpreters in migrant-related settings requires specific language combinations (which also include the European lingua francas) which universities are not always able to cater for. Nevertheless, we feel that this profession is a 'profession of the future' and it is crucial to address issues of training (format and curricula), selection, recruitment, quality and performance now before the current, deeply unsatisfactory, situation in so many countries becomes established as the norm.

The situation we have just described is clearly specific to our own national demographic context and thus both to the history of the profession and the novelty of the discipline locally. The question of broadening or narrowing the field in the classroom will therefore depend on the specific conditions in each country and will affect dilemmas such as: 'which type of terminology to teach and to what degree of specialization (medical, legal, business, etc.)?', 'what aspects of public institutions/organizations should be taught?', 'should the trainer focus more on "straightforward" interpreting or on mediation?, and so forth.

4.1.1 Teaching objectives: interpreting faculties versus modern language faculties

The aim of our course was initially quite simply to train our students to be able to interpret dialogues between English and Italian in the chosen settings. In the interpreting faculty we were working in at the time this suited the faculty's objectives and the students' expectations and competencies perfectly, but for the students we taught subsequently in modern language faculties, where the students have no experience in interpreting or translation, adaptations had to be made to suit their level and objectives. The objective in these cases was to provide an introduction to basic interpreting strategies and communicative skills, as well as providing a theoretical basis that addressed issues of the profession, of language transfer, communication and ethics. With these basic competencies students were able to move on to more technical training courses if they so wished, or to try their hand at interpreting in informal circumstances. In our experience, many students do indeed start working at trade-fairs, for example, without much prior experience and accumulate valuable professional experience there.

In both types of courses (for interpreting students and for modern languages students), we chose the health, legal and commercial settings as the most relevant areas of application for dialogue interpreting (given the need for the migrant population to have access to these services as well as the scope of international trade relations), but we were able to do some work in the area of social services, welfare and education as well and very short incursions into the areas of tourism and the media.

4.1.2   Course format: language-specific or language-generic?

Academic institutions

In an interpreting or language-mediation course, choosing the most suitable language combinations reflecting the needs of the labour market is – in theory – important. Most academic interpreting courses are taught through a specific language combination. See Skaaden and Wattne (2009), for a description of an online module, part online, part on-campus, at Oslo University College for interpreters working in a range of 'minority' languages and covering both theory and practical work. This is an excellent example of how a wide range of minority community languages can be worked into a public service interpreting course. The drawback is that if it is to be implemented in a qualitatively adequate manner, as it is in this case, it requires considerable financial investment. The practical usefulness of the language combination will of course depend on which languages are in demand in the labour market in that particular period. In many tertiary institutions, however, this choice is superfluous, in that the language combinations are dictated by the languages already offered by the faculty, by the students enrolled, or by the language combinations that the teaching staff can offer that particular year. Although there is a certain predictability here that can be ascertained from demographic statistics, these figures change from year to year and thus the needs of the market also change from year to year. This is potentially a major problem for a number of reasons: many of the larger academic institutions (especially the modern language faculties with very large student numbers) are not flexible enough to change the course offered from year to year – indeed it might take years or even decades to accept a new language course in a degree programme and to find and finance the required teaching staff. (Clearly, this situation applies more to the large modern language faculties than the smaller, more flexible specialized interpreting institutions that can more easily introduce new languages, depending again of course on local university policy and budget restrictions.) Some Western modern language faculties that have traditionally only offered courses in the major European languages have started to include languages such as Chinese and Arabic on the assumption that these are the languages of the future, not only representing global economic superpowers but growing ethnic communities in the West. Also, many language faculties are particularly sensitive to the needs of the business community and offer courses in business language, specifying the commercial sector as a potential employment possibility, for example in the international division dealing with foreign clients.

Non-academic institutions

Outside the universities, smaller publicly or privately funded courses that are connected to specific institutions (health, education, social services) are more flexible in structure and therefore better able to meet the language needs of the market at any particular time; they are also of course more susceptible to market needs because their very *raison d'être* and funding depend on the need for communication across languages. (We use 'market' in a very wide sense of the word that covers both the public and private sectors.) Courses organized by municipal, provincial or regional bodies, by non-governmental organizations, or by the institutions themselves (medical, legal, social) really do need to reflect the actual need for interpreters in that particular place and at that particular time. For example, if the institutional aim is to match the needs of the market and if in the last few years there has been a large influx of people from Bangladesh, institutions might choose to invest in the training of Bengali-speaking interpreters if this is at all possible given budget and organizational limitations. Being funded by local, regional or national state-led bodies, and in Europe they might be EU-funded, they are more accountable to their end-users than the larger institutions like universities, which are less directly accountable, at least in the short term, to the general public.

Non language-specific courses

For those non-academic institutions that are more closely in contact with the community interpreting market, a less costly alternative to providing a wide range of binary language combinations is that of providing non language-specific courses. A successful example of a training format that can accommodate a wide variety of languages is an intensive training course for practising, often unqualified, interpreters run by the Norwegian Directorate of Integration and Diversity (IMDi; see Skaaden and Wattne 2009). For these courses, the curricula and the general layout of the programme, the settings and the course progression, trainers could easily adopt the same structure as that of language-specific programmes, but more serious problems emerge in the organization of group work and assessment. One way of getting round this problem is to form groups according to the languages/countries represented by the students and to focus on class group work rather than trainer–student role play in front of the class (see chapter 5). This depends of course on the language distribution in the class.

An example of this particular format was a module in 'language mediation', part of an EU-funded diploma course for cultural mediators of a minority background in Ravenna, Italy, taught by one of the authors. Here, the ethnic mix of the trainees was very varied indeed: Filipino, Moroccan, Ukrainian, Moldavian, Somali, Polish, Russian, Chinese and Albanian. The possibility of forming groups, or even pairs, of the same language combination was not available. One possibility was to use a vehicular lingua franca such as English, French or Russian for role play and interpreting practice, but even that was not always possible. In these cases, especially when the classes are small (not more than twenty), it is best to use the trainer–student role-play format in front of the class (see chapter 5) and have the student summarize/gloss her own interpretation into the host language and ask her to discuss any terminological, cultural or interpersonal problems that arise. It can also be useful to collect information material from public health services or legal services translated from the host language into a variety of languages, and have the class sight-translate these texts back into the host language. In that way the minority language text can be cross-checked against the host language 'target' text. It is also useful to stimulate discussion while presenting the theoretical and technical aspects of interpreting and to get the class members to provide examples from their own languages, leading to discussions using comparative data from perhaps a dozen different languages. Unless there is adequate funding for external examiners for each language, the exam format will have to be adapted to a summary/paraphrase of a text the trainer is familiar with (and knows the contents of) into the host language, or an oral or written exam on theoretical issues discussed in class.

4.1.3   Which languages?

On the basis of immigration statistics one can easily identify the most widely represented national groups in a given country at any given time and, one might assume, those languages in which it would be most useful to train and recruit interpreters. However, even if one were able to offer a series of language combinations that would reflect the current demographic statistics, it is not always easy to predict whether or not the potential users/clients speak the national language rather than a lingua franca like English, French, Russian or Spanish (sometimes for reasons of socio-economic class or professional or regional representation from their home countries), or indeed which national language they are more likely to speak (for example, will a Pakistani migrant in a Western country be more likely to speak Urdu or Punjabi – or both?). A number of other factors complicate this picture even further. For example, many immigrants are bilingual if not trilingual. Generally, but not always, those people who have lived in the country longer will speak the national language better, but this is not always true: there are some communities (especially the women) who prefer to limit contact with the host community and even after many years their mastery of the host language is poor; therefore judging language competence on the basis on the dates of immigration waves (and thus how long people have lived in the host country) does not always yield reliable data. The speed with which a particular immigrant community learns the national language also depends on a number of unpredictable, if not random, variables. (For example, the Albanian or the Romanian community might integrate linguistically more quickly into Italian society than the Chinese community because they hear Italian in their home countries and/or there is affinity between the [Romanian and Italian] languages). The degree of individual integration also varies enormously of course depending on personality, class, education, etc. Other national groups may have a strong community support network of family or friends who function as *de facto* interpreters and liaison units, which limits, one might say undermines, the use of professional interpreters and/or mediators.

Furthermore, not all ethnic groups use institutional services with the same frequency – some are more, or less, represented in the types of crime committed, use the hospital emergency wards more frequently, different types of health problems affect different ethnic communities, and so on. Thus, the health sector, the legal sector and the immigration sector will have different language requirements; regional differences complicate the picture even more. It is not easy to predict, then, which languages will be more in demand in any specific sector (medical, legal, educational) based on general immigration statistics.

4.2   Assessing students' competence

Students' levels of competence and aptitude at the beginning of the course are often extremely varied, and they show a great variety of different skills; these differing levels of competence can easily slow down progress for the whole class. Although assessing students' competence depends on a number of practical factors, such as the number of students, training goals should ideally be identified on the basis of their individual levels of competence, evaluated by a test at the beginning of the course (for example, a language test or a short sight translation into the mother tongue). If the students are aware of their initial level, they can better appreciate the progress made during the course. The level of proficiency required by the course trainers will of course depend on many things, not least on the objectives of the particular institution, which could be training fully fledged interpreters who will then go into the labour market, or which could simply be giving the students a 'taste' of the profession which might lead them to enrol in a more specialized interpreting course. Having ascertained the students' level at the beginning of the course, it is easier to establish, realistically, the objectives in terms of the language level for L2 comprehension and rendition that is expected at the end of the course. For our purposes, L2 comprehension skills are essential. If an interpreter does not understand an utterance and is not able to decipher it from the context with a reasonable degree of confidence, the task at hand becomes very difficult, although other tactics can be resorted to in emergencies (e.g. most obviously, asking the speaker for repetition or clarification). In an assessment phase, students' listening and comprehension skills should therefore be carefully tested, but it is also important to remember that both listening and production skills can be improved in class. Oral production is essential for any category of interpreters, but it is not always easy to assess students' potential in this area. We will not be addressing either assessment or accreditation in this book and refer to the increasing body of data-driven literature in this area (for example, Angelelli 2007; Kainz, Prunc and Schögler 2010).

4.3 Interpreting skills and competencies

The technical skills of interpreting are absolutely fundamental in conference interpreting, and while they are also crucial for dialogue face-to-face interpreters too, we feel we can claim that these skills can be learnt without needing to complete a full three-year degree course, fully aware that this claim is not shared by many of our trainer-colleagues around the world. In those countries where interpreter training simply has not been available until very recently, it is indeed the case that many community interpreters who have learnt on the job, after many years of experience, seem to be able to perform adequately, either because they have a natural talent for this form of interpreting or they learn through experience and practice. However, to be able to deliver the quality performance that each of the interlocutors need and deserve to have, such 'natural talents' must be cultivated, broadened and improved. For a country to be able to rely on a pool of qualified interpreters who can deliver safe and effective quality interpreting, ad hoc solutions, even with the best of intentions and the best 'raw-material', are simply not enough.

Our teaching objective then, as we suggested in chapter 2, is to provide students with a set of interpersonal skills that relate specifically to cross-cultural plurilingual communication which allows them not only to identify and understand cultural and linguistic cues that are crucial to the understanding of the communicative act and its pragmatic function, but also to provide them with what we call transfer skills – how to transfer this pragmatic intent to the listener. If the goal of intercultural communication competence is to implement effective communication strategies that are appropriate to the specific situation, this can only be done by a keen awareness of the cultures at hand and the communication codes embedded in each of these cultures. In the following section we have listed those skills that we believe are essential for interpreter trainees in what we have called the 'a-b-c of interpreting competence'; these apply to both L1 and L2.

4.4 The a-b-c of interpreting competence

4.4.1 Passive skills

A. Language knowledge and comprehension skills
Linguistic level
- *Adequate comprehension of the two languages*: grammar, comprehension, fluency, idiomatic speech, pronunciation. Recognizing rhythm, tone, intonation to identify chunks/signals/turns and to identify irony, sarcasm, humour, metaphorical meanings, implications, allusions, (dis)approval, formality/distance, emotivity, phatic speech markers (performing a social task rather than conveying information), etc. (See dialogues in chapter 6 for examples.)
- *Textual competence*: the comprehension of rules of cohesion and coherence in discourse building; connections between utterances to create meaning by using linking words, reference strings.

Cognitive level
- Memory – ability to remember chunks of speech and units of meaning;
- Ability to capture domain-specific jargon and register;
- Ability to analyse quickly and roughly predict next move to aid memory and fluency.

B. Pragmatic competence
The comprehension of para-linguistic, non-verbal communication (space, gestures, proxemics, body language, tone of voice).

C. Cultural competence

- Recognizing culturally determined assumptions: knowledge of social conventions, institutional practices, taboos and norms;
- Familiarity with institutions: organization of health care and legal services, how to access them;
- Awareness of cultural differences between communication systems, politeness 'systems' and institutional 'communication codes'; recognizing socially and institutionally governed group dynamics (hierarchy, respect, formality, age and gender issues and how these apply in the various professional sectors – at the doctor's, at the police station, in the schoolroom, and so forth);
- Familiarity with professional codes of ethics in various settings (e.g. healthcare, legal) in other countries, if such codes do not exist in the students' own countries.

4.4.2 Active skills: transfer competence and techniques

A. Language knowledge
Language level and verbal reproduction

- The ability to speak sufficiently naturally and idiomatically in L2 (as well as L1) to convey as accurately as possible the speaker's assumed message;
- Reproduction of domain-specific expressions naturally and accurately;
- Combining verbal and non-verbal cues and reproducing them appropriately to the situation;
- Textual competence: the ability to create a coherent text that makes sense to the listener and in which the chunks of speech 'hang together' and sound natural in L2;
- Reproducing language and pragmatic competence features (as above: sarcasm, humour, metaphorical meanings, implications, allusions, (dis)approval, formality/distance, emotivity, etc);
- Reproduction of appropriate synonyms;
- Reproduction, where needed, of repetition and ability to paraphrase adequately, gap-fillers, redundancies;
- Voice-production (clarity, audibility, accent);
- Adequate oral production speed.

Cognitive level

- Memory – the ability to reproduce chunks in units of meaning accurately, speedily, fluently and articulately;

- Decision-making skills: deciding which conversational features are essential and which, if any, are redundant for practical reasons (memory, time).

B. **Interactional skills: group dynamics and interpersonal communication**
  - Situation control/ floor-management and co-ordination – the interpreter as the coordinator of conversation; assertiveness skills (interrupting, asking for clarification of terminology, asserting role boundaries, requesting briefing, requesting breaks);
  - Acting on interlocutor feedback – acknowledging utterances and agreement/disagreement (silence, repetition, gap-fillers, non-verbal cues: nodding, eye-contact);
  - Coping with communication breakdown;
  - Coping with emotivity and tension between the interloctors;
  - Flexibility and multi-tasking;
  - Controlling of bonding and identification with either party: balance between impartiality and positive bonding;
  - Keeping a distance: managing one's own feelings with no outward display of emotions; coping with one's own conflict and emotions;
  - Maintaining confidentiality.

C. **(Inter-)cultural skills**
  - Providing an accurate rendition but at the same time acknowledging the cultural norms of the listener;
  - Rendering accurately and appropriately culture-specific conversation norms (such as greetings, farewells, modes of address, both institutional and social).

Lastly, the interpreter should be aware of her own impact on the communicative event: it is important to acknowledge that the interpreter is not invisible, but an active party in the process of establishing communication. (The literature on the role of the interpreter and her degree of participation in the communicative act is by now vast; Angelelli is one of the key figures in this debate; see, for example, 2004b). It is only through a realistic awareness of this presence that the interpreter can appropriately gauge her involvement without being invasive, thus creating a positive collaborative atmosphere and not intruding on the service provider's area of responsibility. At the same time, it is important that the interpreter does not in any way 'take over' or take upon herself the other interlocutors' roles. To be able to do this she must have an awareness of her role, mandate, tasks, responsibilities and boundaries.

5
In the Classroom

5.1 Structure and organization

Choosing to include the health sector, the legal sector and the business sector as core courses is due, as we explained in the Introduction, to the fact that these are the main areas in which interpreters work in the private and public sectors. They are also appropriate for pedagogical reasons: the variety of terminology and domains adopting the same basic technical interpreting mode provide the student with both a sound anchoring in the necessary skills and a varied repertoire. Including all three sectors would however imply more than could comfortably be included in a short course, if it is to be done thoroughly. Depending of course on the number of hours, this might require sacrificing various items selectively. The basic structure of a course programme can easily be built around the same core features with introductory lessons in a lecture format. Once these basic parameters have been established, the trainer can then move on to more specific terminology and institutional language and register, and lastly to practical exercises in groups. At the beginning of the course students are not usually familiar with technical (especially medical and legal) terminology, even in their mother tongue.

5.1.1 Lesson format – from general to specific to practical exercises

In addition to a general introduction to the profession of interpreting, both internationally and locally, it is useful to provide a few general statistics on the current levels of migration and foreign-language speakers in the country – both permanent residents and tourists, students, people who are in trade, and so on. The official statistics for migration are usually easily accessible and provide the number of migrants, and their ethnic- and language-group breakdown, to give an indication of languages used in the country at any given time – although this changes of course from year to year, as discussed in the previous chapter. Statistics on temporary visitors to the country are harder but not impossible to find (from the Ministry of Tourism, Ministry of Trade, or similar sources). It would be useful to give the students statistics on which particular ethnic groups (and thus languages) access each particular service (hospitals, the police, the courts, welfare offices, job centres, schools) and with what frequency, but we have found that these statistics are rarely available.

The number of lessons on the different institutional healthcare and legal systems in the cultures of the L1 and L2 depends of course on the time available and the expertise of the trainers. Much can be found on the internet, but trainers cannot be expected to be medical or legal experts. Nevertheless, it is often helpful to draw the students' attention to the fact that these differences are significant (e.g. between the Common Law system in the UK and the Roman Law system in Italy) and that often the differences mean that fully corresponding terms may not exist in the other languages. Cross-cultural aspects relating to each institutional field can be taught either in the introductory lessons or when working on that specific sector. It is a good idea to introduce Codes of Ethics/ Standards of Practice at the beginning of the course and then return to these issues after the students have actually tried their hand at interpreting and seen how difficult some of these seemingly straightforward ethical issues can be, after they have seen the videos (see below), and after they have had the opportunity to reflect more carefully about the interpreting process. This brings home more effectively the potential difficulties of full accuracy, floor-management, memory, terminological competence and language fluency.

The material proposed in the course outline given in table 5.1 is ambitious for a short course and whether or not it can be fitted in depends entirely on the number of hours available. For longer courses, however, the trainer could widen the curriculum and add other settings: social services and refugee-related settings, education (schools), tourism, the media, and so on. Once the students have grasped the basic concepts and have had the opportunity to practise in various settings, adding another setting can be done quite easily, through extra reading and terminology lists.

We have also included a lecture on 'varieties of global languages' (for us, that is English): indeed the types of English most often encountered in the legal, health and social service settings are non-UK/US varieties

Table 5.1 Sample of course programme (30 lessons; 60 hours): lesson topics, exercises, activities

Classes	
1–4	**Course introduction** * General introduction to the course – explaining the course structure, reading material, attendance, assessment * **An emerging profession** * An introduction to the history of community interpreting and the state of the art of the profession and the discipline generally: – the development of the discipline within Interpreting Studies; – the 'academicization' and professionalization of practitioners already practising with little or no theoretical training. **An introduction to interpreting for private and public institutions in the country at issue** * Differences in terminology: dialogue interpreter versus community interpreter versus public service interpreter, and conference interpreting versus dialogue interpreting: differences in context and setting, differences in techniques; * How the relevant institutions are organized in your country; * The current state of migration: immigration statistics and language communities in the host country; * Issues of cross-cultural communication: living in an ethnically and linguistically complex community; * A brief introduction to codes of ethics and standards of practice: general parameters.
5–6	**Interpreting skills, competencies and techniques** * Practical interpersonal communication issues: interrupting to ask for repetition; interpreting in the first person versus the third person; the importance of politeness strategies in casual and professional/institutional settings (professional and social hierarchies, age, gender); greetings and terms of address; speaking in public: articulation, voice, accent, posture; interpreting gap-fillers, hedging, phatic language; the importance of implicature, irony, humour, affect (anger, tension, emotions); body language; stress factors (anxiety, nervousness and stress, vicarious trauma; acoustic and other practical problems). * Logistics (how do we sit? facing the interlocutors? how does this affect logistics and interpersonal rapport?); * Completeness: how much of the utterance should be interpreted? * Interpreter skills and competencies – what is required of an interpreter? (language competence versus cultural and interpersonal competence); * Theoretical implications: to what extent is the metaphor of the 'interpreter's invisibility' correct?

(*continued*)

92

Table 5.1 Continued

7	* 'Breaking the ice' games and exercises; * Memory practice, sight translation, listening practice (BBC, CNN, YouTube, non-standard accents). * Note-taking.
8	**Interpreting for the business sector** * Models of cultural difference, ethical considerations, examples of culture differences and the impact of interpreting on business negotiations; * Discussion on the particular requirements of the business sector and differences between private and public settings.
9–11	**Business terminology, role play and dialogues** * Interpreting a simple dialogue from the business sector with teacher–student role play and the rest of the class as audience (peer feedback and trainer's comments); * Discussing examples of mixed registers – especially pertinent in the business sector (moving between conversational discourse and professional/technical discourse); * Role play, acting out dialogues.
12	**Interpreting for the health services** * Models of cultural difference: ethical considerations; * Institutions and practices in the home country - a brief description of the health system and principal actors in the two countries.
13	**Terminology in healthcare interpreting: role play and dialogues** * Basic terminology; * Interpreting a simple dialogue from the healthcare sector with teacher–student role play and the rest of the class as audience (peer feedback and trainer's comments).
14–16	**Other exercises** * Student assignment: choose a situation and work in groups to prepare setting, terminology and 'scenes' for following week; * Some sight translation and memory exercises.
17	**Mental health interpreting** * Case studies; * Video on mental health interpreting.
18	**Interpreting in the legal sector** * Institutional and cross-cultural aspects; brief description of the principal actors; * The courts, the police; * Basic legal terminology.
19–21	**Practical exercises in the legal sector** * Interpreting/memory practice with simple legal texts; * Interpreting a simple legal dialogue with teacher–student role play and the rest of the class as audience.

(continued)

Table 5.1 Continued

22	**Video on legal interpreting** Discussing ethical issues: confidentiality and neutrality
23–25	**Codes of ethics and the interpreter's role** * Looking more in depth at the field-specific aspects of various codes of ethics and discuss general theoretical questions: what is the ultimate goal of the communicative act – communicate come what may? What does a service provider expect? Alignment and taking sides, the institution's expectations; * Discuss students' experiences and thoughts in the light of interpreter ethics after having practised for a whole semester.
26–27	**A wider framework: how culture shapes the way we speak. Issues of cross-cultural and intercultural communication models** * Models of intercultural communication. 'How culture shapes the way we speak: the importance of culture in communication models – do we really understand each other? * Discussing group dynamics, interpersonal interactions, the interpreter as 'cultural expert', subjectivity and the interpreter's own culture (the subject always looks through the filter of his own culture); * Hybrid identities – especially related to L2 culture (for English: Asian or African diasporas; cultural and linguistic aspects).
27–28	**Varieties of English (or other L2 lingua francas)** * Comprehension and text-reproducing exercises: practising listening to non-UK/US forms of English (CNN, BBC World, YouTube): terminological/syntactical and non-verbal differences (e.g. Indian-English, Nigerian-English)
29–30	**Summing up and on-the-job issues** * Interpreter self-confidence and self-protection, relaxation techniques, how to keep a distance from topics/actors; * Job opportunities; * Exam practice.

of English (for example Nigerian English) or English used by Japanese, Chinese or Arab speakers in the business world. Even just a bit of listening practice is enough to sensitize the students to this issue; further 'listening and reproducing' practice can be done at home or in the language laboratory from the major television channels (for English L2: BBC World and CNN, YouTube).

This outline in table 5.1 is thus meant to be a blueprint for trainers from which to extrapolate (or add to) material they find useful for their students and relevant to the training institution as well as to the country and language(s) in which they are operating. We integrate salient aspects of cross-cultural communication into all our lessons and role plays, but because we believe intercultural features in our profession are so important – and fascinating – we try to include a lesson on more theoretical aspects of culture, intercultural communication and cultural identity. Whether or not to include these aspects will of course depend on the trainer's interest, inclination and expertise, as well as time.

Ideally, theory should be carefully balanced with various types of practical exercises either in alternative lessons or in the same lesson. However, students' attention span decreases (especially if the lesson is longer than 60 minutes) and practical exercises require organization. The number of students attending lessons also has an impact on the opportunity to practise. A class of 20 or 25 students allows them both to work in small groups and do role plays with the trainers. Thus, they are able to 'perform' several times during the course of the module; this will of course depend on class size and number of lessons.

5.1.2   Role play and pre-prepared dialogue simulation

We use two basic formats for role play and dialogue simulation, alternating in class between these two different kinds of simulation exercises. The first format is that the teacher(s) and one or more volunteer student(s) act out the dialogue at the front of the classroom with the rest of the class looking on. One student at a time is asked to come to the front of the room and interpret as the trainer or trainers read and play out a pre-prepared dialogue in the two languages. The other way of acting out simulated dialogues is to divide the class into groups (three to five students, or more if the classes are big) and to distribute a pre-prepared dialogue to each group, which can be circulated to other groups when they have finished.

The scripts should be read carefully with clear articulation and pronunciation, especially at the beginning of the course when the students are not yet used to the individual trainer's voice, accent and communication style. The importance of intonation for comprehension is often underestimated, not least for non-native language students listening to a native language speaker speaking quickly and fluently, and this should be taken into account. As the course progresses and if and when the students are able to meet the challenge, trainers can deliberately read quickly or less clearly to increase the level of difficulty and to simulate real-life situations. At the end of the dialogue, the student who is interpreting is then asked to tell the class and the trainer(s) how she felt, what was most difficult, where she felt she went wrong or did particularly well, her level of stress and any other aspects she might choose to comment on. The other students are asked to observe their fellow-student's performance during the session, assessing features such as translation accuracy, strategies employed to deal with interpersonal or cultural communication problems, and so forth, and comment on these afterwards.

Role play in groups

Role play in groups can be either planned or unplanned. Each student will then choose a role in the dialogue: for example, one student will choose to be the service provider (doctor, patient, nurse, teacher, parent, lawyer, judge) and one will choose to be the person who does not speak the majority language; if the class is big and there are more than three or four students in each group, then they might take turns interpreting or playing out the different parts, or in using their imagination to invent a variety of supporting roles. After taking five minutes to read through the dialogue (not the student who will be acting as interpreter), the students can then act it out themselves. This way, the students actually get to practise interpreting in all its facets – neuro-linguistic, language mastery, terminological (field-specific and general), as floor coordinators managing the interpersonal rapport between interlocutors, managing social, cultural and institutional aspects of the communicative act, and dealing with practical and logistical issues such as poor acoustics. Bringing a portable electronic dictionary, or online glossaries, to class helps solve many of the immediate terminological translation problems.

This group exercise is far less daunting for most students than having to perform in front of a class, and it also fosters class cohesion. We will usually 'patrol' the classroom and listen in to the dialogues, give suggestions, answer questions, and so on. Having two trainers is not essential when using this format, but it does of course help to have a native speaker in each of the two languages. We usually give the students 15–20 minutes to play out each dialogue and then ask them to circulate the dialogues and try a new one. Depending on the size of the room, the size of the class and the students' capacity for concentration, this exercise can sometimes end in private conversations between the students and they may need a bit of prodding to focus on the task at hand.

This format clearly has the advantage of involving all of the students so that by the end of term each student has been able to interpret several times in several different settings. Also, it tends to foster a good rapport between the students, and they feel freer to explore the interpersonal dimensions of group dynamics between the various 'characters'. Ideally, the trainer should guide the groups as they act out their role plays, to address the theoretical issues discussed at the beginning of the lesson. We found, however, that group work is often unpredictable. The drawback is clearly that the trainer cannot fully monitor the situation in each group and that the students cannot observe the performance of their peers. Class discussions rarely ensue in these cases, unless there is time for one of the groups to 'perform' in front of the class at the end of the lesson. With small classes this is easier of course, and the ideal situation would be one in which each group met during the week, prepared a dialogue (not shown to the chosen interpreter) and acted it out in front of the class.

Team-work is generally a good way to start, a sort of breaking-the-ice exercise, as students tend to be more confident if they are working with colleagues and they 'share the burden' of their early interpreting tasks before they 'perform' in front of the class. Indeed, one of the main difficulties they initially find is that of speaking in front of an audience. Addressing practical issues related to public speaking, even cursorily, is a good way to boost their self-confidence. It also allows the students to ask questions, voice their fears and share these with their peers, taking some of the initial sting out of the first couple of performances. Some students are 'naturals' and enjoy the attention, and there is also the danger of allowing these students to 'take over' and monopolize the sessions. Usually, by the end of the module, even the most timid students come to enjoy the role play.

5.1.3   Preparing the dialogue at home

Using pre-prepared scripts is by far the most time-effective and easiest way to organize role-play techniques, and at least once or twice per semester we ask the students to write their own dialogues, doing terminological research at home to prepare for the following lesson, and encourage them to start preparing a glossary. These need not be real scripts, but rather notes or a schematic outline describing a situation, with the relevant terminology they can improvise from and a variety of roles to practise (in the business sector: Marketing Director, the General Manager, and the would-be car salesman). On other occasions we have asked the students to prepare a topic at home or improvise a dialogue, by giving them a specific situation (e.g. 'an Italian tourist in London has broken his leg and has just been admitted to the emergency ward; he does not speak a word of English and the staff interpreter has been called in' or 'an Australian businessman is rushed to hospital in Tokyo').[1] Group work is done in class the following week and an interpreter is chosen from a different group just before role play starts. Preparing at home may be logistically difficult (if students find it difficult to meet after class), but it is an excellent exercise to get them to practise doing terminological research work, to navigate the web creatively and to learn how to choose original material and not translations of corporate websites. It also stimulates their imagination and makes them more responsible for their own performances.

The success of these exercises, in our experience, rests in part on the dynamics between the students and they are usually more successful in a small class (up to 25–30 students) than in a large one. There have been times when we have felt enormously privileged when the class 'chemistry' between the students as a group and between them and us has been just perfect. We remember especially one particularly imaginative group who, to their peers' amusement and the trainers' astonishment, wheeled a 'patient' in a wheelchair into the classroom to the sound-track of the Brazilian soap-opera '*Beautiful*' that was being broadcast on national television at the time and was hugely popular. In a perfect send-up, they then proceeded to act out a scene where the main character 'Ridge' had broken his leg and was taken to hospital in a foreign country and needed the services of an interpreter. The terminological training might not have been particularly sophisticated, but it did a great deal for class cohesion.

In the legal setting, introducing 'softer' pedagogical strategies can sometimes be useful too, such as watching courtroom dramas ('*My cousin Vinny*') that provide a wealth of information on terminology and institutional and interpersonal relations. In one very successful exercise students were asked to conduct a mock trial based on the television series that they had watched. They were divided into teams of six or seven and provided with background material about a case that they had to read for the following lesson. Each group member was assigned a role and each group was asked to create a mock court case.

5.1.4 Trainer–single student role play

The major disadvantage to this role-play format, that is the 'trainer(s) plus student' pair in front of the class, is time. In a one-hour lesson, three or four students at most will be able to perform. The advantage is that the trainer can monitor at all times what is happening and can correct and comment continuously. The class as a group can observe how the actual interpreting situation is played out on the ground and can observe and comment on good practice and errors. Good group discussions between the whole class are fostered this way, sometimes dwelling on seemingly trivial semantic or interpersonal details that lead to productive discussions on various translation and interpreting features. At the beginning of the semester students are often shy and unwilling to perform in front of the class, especially if they have never interpreted before. Once they realize that most students find this task difficult and that even the best students make errors, they feel more self-confident. A good way to start is to get the students used to interpreting gently through memory exercises, sight translation and other 'ice-breakers'. Those students who have not previously had any interpreter training at all might benefit from preliminary exercises such as short sight translations and memory tests to train their reactive competence and to gently ease them into the more challenging interpreting tasks. This preparatory phase can be controlled and guided by the trainers adapting both pace and terminology level to the individual student's abilities as they emerge during the semester.

There are many ways of helping the students as they play out the dialogues. If they get stuck for a word, trainers might ask pertinent questions as the pretend interlocutor: 'well?', 'and then what happened?', 'and?', 'I'm not sure I understand?', 'Oh', etc. When students ask for repetition, they should be encouraged to stay focussed on what is happening in the dialogue (it's easy to get distracted because of the stress level), to use their imagination and envisage that they are in that situation. If the students make a mistake, the trainers can help them by asking 'I didn't hear the number of orders?', '...the date?', 'the name?', 'the last item in that list?' (which is also how the students should be taught to ask for repetition – by asking specific questions, not just 'I forgot what you said'). A minimum of imagination – i.e. imagining oneself in that situation – coupled with a high level of concentration will also help the student understand the logic of the sentence. This is important because logic also aids *memory* in the comprehension phase and thus aids understanding and accuracy in the reproducing phase. It's also good to remind the students that the more self-assured they appear, the more their interlocutors will trust them as convincing professionals and by that same token trust the accuracy of their rendition.[2] We remind the students that memory is a complex, interrelated system that involves numerous neurological and cognitive processes and comprises various phases. The *encoding* phase is stimulated by visual or aural input, and processed at the linguistic level through semantic meaning, grammatical rules and pragmatic function, but it is also mediated through personal experience and past memory. Efficient *storage* of information comes with practice, and allows the brain to store only that information which is necessary, either in the short or long term.

Interpreter underperformance due to *retrieval* problems (forgetting) can be caused by lack of attention/concentration, lack of comprehension (successful cognitive or aural encoding) or even simple distraction due to poor acoustics or other practical problems.

5.1.5 Recording student deliveries

If possible, recording the students' deliveries and commenting on them afterwards, especially if there is any doubt about the rendition, is useful. Today, with the help of digital recorders, recording the students is unobtrusive and simple: one of these small devices can simply be put on the trainer's desk or on a desk or a chair between the students when they are working in groups. If on the trainer's desk, there is no need to move it or put it close to the interlocutors – they can be positioned at quite a distance if the *zoom* mode is set, and after a while students are no longer even aware of being recorded. Digital recorders can be connected to a computer, the recordings unloaded and copied onto a USB-drive, so that students can hear their own renditions at home and analyse them. We always encourage them to work with a partner, and compare their renditions. This usually makes them feel more self-confident and encourages them to be even more self-critical. (An initial investment is needed, but prices of electronic equipment are no longer prohibitive.)

5.1.6 Evaluating delivery through peer critique and self-assessment

We have found that having to formulate comments as peer critique or self-criticism in response to their own or their classmates' deliveries is challenging, but useful. Because interpreting can be such an anxiety-provoking activity, it is important to strike the right balance between reassurance and constructive criticism. Trainer critique is invaluable of course, but as the students gain confidence in their own abilities, as their terminological competence improves, and as the class gradually bonds as a learning unit, peer critique is perhaps just as valuable, because it creates a deeper self-awareness about both technical issues (accuracy, speed and clarity of delivery, turn-taking techniques, mnemonic needs and limitations and the need to interrupt, voice, accent, terminology, register, connotation, translation, politeness, posture, eye-contact) and wider cultural or conceptual issues.

Evaluation is always subjective to some degree: instead of, quality is notoriously difficult to quantify, and critique can be potentially highly face-threatening; but if the students have a standard set of broad parameters that they are all using, it can be done effectively and gently. General parameters might be:

- Was she clear and accurate?
- Did she make any contradictions in terms?
- Was (s)he fluent, or hesitant in L2?
- Was the pronunciation acceptable?
- Did she use the appropriate register and terminology?
- Were there any significant omissions?
- Was eye-contact, body positioning/posture, tone of voice appropriate to the culture and/or context?

At the beginning of the course students tend to be excessively sympathetic vis-à-vis their classmates and are reluctant to point out errors, but when they understand it is both in their own and their classmates' interest, and that they are all applying the same parameters, they feel more comfortable about collaborating. Gradually, they become increasingly aware of what can be expected from an interpreter, the main skills to be acquired and improved, and problem-solving strategies. Even the very fact of being aware that the same or very similar difficulties are shared by their classmates can be encouraging. The ensuing discussions in class are often both interesting and enlightening, as comments allow students to engage actively in analysis of all the linguistic and cross-cultural issues that emerge during the simulation.

5.1.7 Evaluating accuracy in student delivery

Evaluating quality in interpreting performance is notoriously difficult and subjective, as is testified to by the vast quantity of literature on quality in translation and conference interpreting. The emerging literature in community interpreting and public service interpreting referred to in chapter 1 is beginning to address these issues rigorously with data-driven scientific analysis, and we can only refer the reader to these works. However, since this book is meant to be a practical guide based on our own teaching experience, we would nevertheless like to suggest a few practical criteria that trainers may find helpful.

One simple way of formulating interpreter performance that employs reasonably quantifiable features is to use Bente Jacobsen's categorizations of interpreter additions[3] of varying significance, which has been useful to us (Jacobsen 2002). Using the criteria of 'additions', she distinguishes between additions of no impact, of minimal impact and of significant impact, as we can see in table 5.2. There will always be a

Table 5.2 Jacobsen's model: the main categories of interpreters' additions (Jacobsen 2002: 241–2)

Additions with no impact on the semantic and/or pragmatic content of the source text	Repetitions	
	Silent pauses	
	Voice-filled pauses	
	False starts	
Additions with minimal impact on the semantic and/or pragmatic content of the source text	Repetitions	
	Fillers	
	Paralinguistics	
	Explicating additions	Obvious-information additions
		Connective additions
		Additions explicating culture-bound information
	Elaborating additions	
Additions with significant impact on the semantic and/or pragmatic content of the source text	Emphasizing additions	
	Down-toning additions	
	New-information additions	

measure of subjectivity in assigning value to these language features (the degree of impact), but nonetheless such additions are easily iden tifiable.

Additions with a significant impact on the semantic and/or pragmatic content of the source text, categorized as 'brand-new', i.e. explicitly or implicitly introduced into the interaction for the first time, would be considered errors (ranging in degree from emphasis to addition of new material), unless the student can argue that it is necessary for global accuracy. These can be considered *emphasizing additions* which serve to emphasize or increase the force of the original utterance, either explicitly or implicitly, or by adding stress to one or more items in her target text. The opposite, *down-toning additions* attenuate or decrease the force of the original utterance. Finally, *new-information additions*, Jacobsen says, neither emphasize nor attenuate the force of the original utterance, but nevertheless introduce brand-new information (Jacobsen

2002, adapted from Garzone and Rudvin 2003). The other two categories, additions with no impact or with minimal impact, present more nuanced 'errors', but can lead to fascinating discussions on the nature of translation, on vocabulary, cohesion, pragmatics and the language/culture bond between whatever happens to be the group's L1–L2 combination. Indeed these factors are excellent opportunities for improving the students' mastery of L2 (traditional 'language practice'). Sometimes they also uncover differences in opinion and raise discussions in class on the nuances of lexical items and other features of the L1.

This schema can be used for peer criticism as well; it is useful because it gives the students an evaluation instrument which is both tangible and user-friendly and, by specifying a particular category of rendition or addition, is less face-threatening than simply judging the rendition as 'inaccurate', 'bad', 'sounds wrong' etc. It is also important to get the students to practise identifying interpreter errors and to reflect more carefully on their own performance.

5.1.8   Trainer–student rapport: the advantages of two trainers versus one

Although this is not always possible, we have found that the ideal teaching situation is when two trainers are present at the same time – one for each language. The advantages are obvious in that students hear and get language feedback from native speakers in each of the two languages. Of course, the major disadvantage here is cost, and for most institutions this approach is beyond the limit of their budget.

However, the advantages of having two trainers are worth the investment: not only can each trainer comment on the specific problems in each language (both technical and relating to everyday language) and give further suggestions, but it is much easier to act out in a plausible manner the simulated dialogues and to guide the students to include whatever particular issue or problem is to be addressed during that particular lesson. When two trainers collaborate, dialogues can be prepared much more accurately, as each teacher brings in their personal, professional and teaching experience and can touch upon multi-faceted aspects of cross-cultural issues, especially if the trainer has a previous and/or ongoing migratory experience or recalls early difficulties in the host country. It is useful for the students to observe the interpersonal dynamics as they unfold between two interlocutors who are in control of the situation. Furthermore, students' renditions can be analysed much more thoroughly, as in-depth comments on the students' solutions, concerning syntax, grammar and terminology can be given in both languages.

Another advantage to this format is that teachers can improvise. It is in fact always easier for students to interpret interlocutors who are speaking ad lib and who are involved in the dialogue, while if they are reading a text they tend to be quite detached (from the subject-matter) and speak fast. Moreover, such simulated and partly improvised two-interlocutor dialogues are much closer to real-life situations. For example, Dialogue 11 in chapter 6 is based on the real experience of a colleague who needed to renew his residence permit. When acting it out, it looked like a real impromptu dialogue. Of course, teachers are neither real actors nor are they expected to perform as such, but our experience shows that students tend to perform better in these conditions.

The dynamics that emerge between the two trainers as they act out the dialogue along with the student-interpreter is thus significantly more beneficial than in a one-trainer situation. This is due perhaps to the individual personalities of the trainers but also, as mentioned, to the fact that they will inevitably have different professional experiences and will bring these to bear on the role play. The combination of these professional and life-experiences can lead to lively discussions between the two trainers and between the trainers and the class. In our case, the combination of a more theoretical experience with a more practical one was, we feel, successful. The discussions that ensued between us and the students gave them far more, we believe, than if they had only had one of us available at a time.

5.1.9 Writing dialogues

Ideally, dialogues should be based on real situations, the teacher's personal or professional experience or that of her colleagues. Some of our business dialogues were based on the early professional experience acquired mainly in trade-fair assignments in Bologna. Another frequent real-life situation is that of migrant workers, for example applying for a residence permit. Again Dialogue 11 in chapter 6 gave us the opportunity to carry out terminology work (e.g. residence permit, tax stamp, proof of financial responsibility, etc.) as well as to tackle cross-cultural issues with the students. However, many good dialogues are also pure invention (indeed, carefully scripted dialogues can be easier to adapt to the specific pedagogical needs of the students). Sources of inspiration are infinite, from newspapers, magazines, television, language textbooks, the internet, etc. The internet is a useful source for witness reports, post-mortem and cross-examination reports. In the medical setting there is an endless amount of material on the websites of cancer survivors, medical societies, weight watchers, hospitals transplant centres and so forth. Dialogues are also good for expressing a range of registers in one and the same situation – from technical language to disjointed natural conversational speech and give the students a reasonably accurate idea of what they can expect in a real-life situation. What should a dialogue contain? Ideally, most or all of the various phases of a conversation, that will vary from field to field – where certain regular patterns are distinguishable – and, obviously, from situation to situation. As a rule of thumb, we would say that a dialogue should contain: greetings, introductions, small talk, dates, names, figures, places and both general as well as field-specific language and technical terms. Introducing cultural aspects (food, historical sites) is also a good idea, as well as the 'obstacles' we mention below

Wire-tapping: Interpreters working for the police are often called to translate intercepted telephone calls, and these are notoriously difficult. We have not included this format in chapter 6 because they are best rendered on a CD in order to maintain the acoustic and language obstacles: strong accent/dialect, overlapping speech, mumbling, inarticulateness, slang and coded slang, and so on. But trainers could easily write and act out, or record and play in class, imagined telephone conversations; asking the students to transcribe and translate them is even more useful as a pedagogical exercise.

5.1.10   Examples of cross-cultural pragmatic features that can be built into dialogues

Incorporating politeness, 'Ps and Qs'

We cannot stress enough how complex and how prone to misunderstanding politeness strategies are cross-culturally. The most emblematic example of this is perhaps the politeness system of the English, where reserve, embarrassment and extreme politeness interconnect in a complex communication network often completely unintelligible for the outsider, and open to constant misinterpretation.

Situations containing the exchange of goods and services are interesting for interpersonal dynamics. Asking for something can be a potentially face-threatening situation and must be negotiated very carefully so as not to impose or offend. Hedging, indirectness, forming requests indirectly in the declarative or interrogative form, and so on, are tactics used in most societies. Using the imperative to give an order can easily cause offence, as might a direct request. It is good to get the students used to changing not just syntax but mood by remembering that the overall objective of translation is (generally) to maintain the communicative function.

Other aspects of politeness are just as complex. Note how 'please', 'thank you' 'excuse me', 'pardon', 'I beg your pardon', depending on context and tone of voice, are not always as 'innocent' as they seem, but may contain all kinds of reticence, irony and even rudeness that must be read between the lines. (Note also 'Excuse me!' in a tense situation as a rebuke.) Egalitarianism and respect might be implied on the surface, but lurking underneath there might be a hotbed of other objectives. Aggressive politeness can create distance and asymmetry, putting the interlocutor 'in place', and 'frosty' politeness can be used to express displeasure. These are excellent cross-cultural features, that potentially create huge misunderstandings, to work into dialogues.

Sorry
'Sorry' is another interesting politeness marker, in that it can be both an apology and a discourse marker: 'Sorry, could you do this for me?', 'Sorry, could you get me that?', 'Sorry, we agreed on 15,000', 'Sorry, I'm not taking no for an answer'. There is much potential for cross-cultural confusion that can be worked into dialogues here. (See Fox 2004: 148 for an excellent discussion on the English ubiquitous and often counterintuitive use of 'sorry'.)

Consider the following (B is the English speaker):

A. The price is 120,000
B. Sorry, I thought it was 100,000 (the interpreter uses a word that indicates a real apology)
A. That's OK, no need to apologize, it's 120,000
B. Sorry, last week you said 100,000

Here we have a clear misunderstanding that the interpreter needs to clarify.

Other cross-cultural pragmatic features
A number of more mundane, cross-culturally interesting, dialogues can be built around everyday occurrences:

Punctuality
(A is the English speaker)
A: I thought we agreed to meet at 9 o'clock this morning?

B: That's right, it's nine-thirty. Did you sleep well?
A: I hope this isn't the way you do business...
B: Yes, let's talk business now...
A: Ok (rolls his eyes); that's exactly what we've been doing for the last half hour. Now we can start at last. Thanks.
B: No problem, I'm really happy to talk to you

Pain
Expressions of pain can also lead to misunderstandings ('Does it hurt?', 'How much?', 'Just a little', while grimacing with pain).

Attitudes
Different notions of political correctness can easily be worked into dialogues, (especially towards women as secretaries, women doing the housework, women drivers, etc.). Adding humour and irony to dialogues can be a very good way to foster good class relations.

Hyperbole and indirectness versus facts and figures
The tendency to exaggerate, use hyperbole and be indirect is often frowned upon in more literal-minded cultures or more literal-minded contexts, such as the police station or the courtroom.

First name or title and surname?
The use of first name versus surname may cause great embarrassment and this too could be successfully worked into a dialogue. A British, American or Australian manager might address an employee using his first name. If the employee does not come from northern Europe, the US or Australia, it is quite likely that he would feel very embarrassed by this. Note also the tendency of North American speakers to use the interlocutor's first name repeatedly at the beginning of a sentence, usually as a vocative, or in mid-sentence. This also applies to self-introductions (for example waiters in restaurants, an increasingly common trend).

   We also find modern colloquialisms that are so easily misunderstood:

An Italian engineer, Paolo, is working for a company in the US; the female manager, Miss Evans, is American. There has been a delay with the delivery of some pipes (E = English speaker, IT = Italian speaker):
E: Paolo, you wanted to see me
IT: uh…, yes (*avoids using her first name and seems evasive*).
It's about the tubes.
E: Tubes?! I don't remember any tubes.
IT: I'm sorry, you said you wanted to talk about tubes for the road works.

E: Ah, you mean pipes? We call those pipes in the US.

IT: Excuse me, I used the wrong word. <u>You see, there's a little problem.</u>

E: <u>Are we talking about a big problem, or a small problem here,</u> Paolo? We need to address this issue.

IT: No **problem**, Miss Evans, just a small delay.

E: <u>Are we talking days or weeks here?</u>

IT: Talking?

E: I mean, will it take a few days or a few weeks?

IT: Just a few days.

E: Paolo, We have to have it done by Friday.

IT: <u>Oh, no, Friday, that's impossible.</u>

E: <u>We don't use the word impossible in our firm, Paolo.</u>

IT: Eh?

E: <u>Paolo, those pipes have to be in the factory by Friday and in place by next Wednesday. Do we understand each other?</u>

IT: Yes, I understand, Friday …. (and a new possibility is introduced for further misunderstanding….)

We have not found many good sources for dialogues on the internet (although interviews are an exception as they already work as dialogues if one translates one of the speakers). Sandra Hale (1996) has a collection of 38 medical and legal Spanish–English dialogues that are excellent for classroom use, available on http://www.eric.ed.gov/PDFS/ED424759.pdf.

There are also a number of websites, and many simple textbooks for language learners, that provide examples of greeting and leave-taking in different languages. http://kkitao.e-learning-server.com/corpus/function.doc provides a simple list of English expressions for the various phases of conversation (greetings and leave-takings, apologizing, introductions, making conversation, invitations, compliments, feedback, making complaints and suggestions, giving advice, correcting, expressing appreciation, congratulating and expressing sympathy) and various ways to request and exchange goods and services (making requests). See also Knuf (1990) for a bibliography of greetings and leave-takings. Given that these seemingly innocuous conversation norms can actually be a recipe for disastrous social gaffes, it is time invested wisely – for trainers and students – to do a bit of research in politeness strategies in L2 and perhaps even compile a glossary of stock phrases.

5.1.11   Making life even more difficult

Ideally, dialogues should become progressively more difficult. The level of difficulty can be increased in a variety of ways, from the perspective of terminology and technique to interpersonal dynamics. As the course progresses, it can be useful to introduce stress factors to the role play, simulating a range of real-life situations that the students are likely to encounter (for example, in the context of English, using non-standard accents, e.g. not just mother-tongue Indian or African English but a Japanese businessman speaking unsteady English). Difficulties may arise from the translation of jokes, especially when they include puns that cannot be literally translated into the other languages. In business settings, for instance, interlocutors may start with a joke, just to be friendly, but jokes are often untranslatable or they may not be funny at all once they are translated, or one of the interlocutors may feel offended and may react to the other party's joke or comment negatively.

Language and practical obstacles

- Acoustic problems;
- Confusing para-verbal elements such as gestures, body language, silences;
- Poor articulation (speaking softly and/or indistinctly);
- Confused sentence formulation;
- Difficult terminology for which the students are unprepared (which forces them to paraphrase and ask for clarification from the speaker);
- Extreme variations in register (for example a paediatric consultation with register varieties from baby-talk to medical technicalities, or a business meeting spanning introductions and pleasantries to technical discourse);
- Confusing and misleading kinship terms as identification parameters ('brother', 'sister', 'cousin' or even 'mother', 'father', 'grandfather', 'auntie', 'uncle').

Interpersonal communication obstacles:

- Strict institutional hierarchy negatively affecting rapport, language register and politeness;
- Constant interruptions by speakers;
- Overlapping speech;
- Failure to address the interlocutors in an appropriate manner;
- Not allowing for pauses and respecting interlocutors' silences;
- Tension among the interlocutors; anger; vulnerability and fear of interlocutors (prison, police station, operating theatre); aggression; affective stress factors can also be introduced – anger, pain, fear, hysteria, arguments, aggression, hostility, emotion (crying);

- Turning to the interpreter for assistance – or the interpreter blamed when things go wrong;
- Socio-cultural taboos, swear words, sexually explicit language;
- Socio-cultural stress factors such as disparities in age, gender and ethnicity;
- Ethical dilemmas (torn between two parties).

5.1.12   Authenticity

Although using authentic material (recorded dialogues of authentic speech) is desirable, we found that because of the difficulty of accessing such material and because authentic material is more difficult to tailor-make for pedagogical purposes, constructed, simulated dialogues were perfectly adequate for our purposes. When two trainers are present, dialogues can easily be improvised into a plausible rendition of real-life situations. It is important to be able to construct the dialogues so that they include all the elements one would like to address (greetings, welcomes, general and technical terminology). These elements may be specific to a particular culture or language combination and they must be adapted to suit the level of the students.

Students get a good deal of exposure to authentic material through listening and memory exercises, however, and through the exercises they are encouraged to do at home: practising summarizing, paraphrasing and interpreting while listening to the radio and television (see section 5.2). This is not as effective as role play, but it does give them ample exposure to pertinent text types in L2.

5.1.13   Guest speakers – lectures and role play

Inviting service providers or academics (lawyers, magistrates, doctors, lecturers) is always an added bonus. A lecture on the role and legal liability of interpreters working in the legal or medical setting by a legal expert, is one example. Comparative analyses of different legal systems, tackling basic terminology-related issues, describing cases in which interpreters were convicted for malpractice or held responsible for misdiagnoses are all interesting possibilities. The ideal situation is not to overwhelm students with excessively technical information, but rather to make them aware of institution- and system-related aspects of the various interpreting settings and the interpreter's accountability. Our invited speakers on these occasions in the legal sector have ranged from police and court interpreters, police officers/administrators to lawyers. Similarly, for the health sector lectures, we have invited doctors, nurses, physiotherapists and hospital interpreters to talk about work involving cross-cultural communication. Being at the point in their lives when they are about to decide which career path to choose, students find it particularly useful to hear about the everyday work of interpreters and to imagine themselves in that same situation.

If the speakers agree, then real-life interpreted situations can also be simulated with their help: a doctor or nurse from a local hospital offers an excellent role-play opportunity. In the post-graduate intensive course on public service interpreting, for example, we invited an interpreter colleague from the hospital in Rimini – one of the few Italian hospitals that supply an internal interpreting service. She was able to describe real-life situations accurately and vividly and to simulate doctor–patient dialogues realistically; students were called upon to interpret these dialogues and were assessed by an outside experienced professional. Her presentation also touched upon hospital organization, how to interact with healthcare professionals, work safety issues, as well as all the other tasks that an interpreter may be required to fulfil in that setting, namely written translations of test outcomes, interpreting on the phone for foreign insurance companies or specialists, helping migrants with bureaucratic procedures, and so forth, She also discussed cross-cultural and ethical issues such as safeguarding the patient without taking sides or being considered as an 'ally', acting as cultural clarifier when physicians use medical jargon and patients do not understand them. Real-life practical testimonies such as these are both motivating and educational. Students from other relevant Faculties such as Law, Medicine/ Nursing or Psychology – see Skaaden and Wattne (2009) – can also be involved in role plays to make the activity both more realistic and more interdisciplinary and this could lead to a profitable mutual learning experience.

This type of teaching format is particularly useful as it not only gives the students a very clear idea of what goes on 'on the ground', but it allows both speaker and teacher to touch upon theoretical and practical issues at the same time, illustrating one through the other. It is a perfect opportunity to discuss various aspects of the code of conduct and to problematize these issues by referring to real-life situations and real-life people working in the field and with a vested interest in optimizing communication. An additional benefit is that it helps strengthen the bond between the training institution and professional institutions. It also creates a direct dialogue between students and service providers, and through this dialogue the students feel that they are contributing towards 'educating the service providers', a need that has been pointed out time and time again in the literature on interpreting. Time and circumstances permitting, we also included visits with the sutdents outside the classroom, for example to the City Court, or to conferences and workshops.

5.2 The basics of floor-management in the classroom: memory, turn-taking and interruption strategies

Discourse coordination is realized primarily by turn-taking in which turn lengths may vary from a one-word utterance to a whole speech/story. There are no hard and fast rules determining the order of turns, but there are discoursal rules for transition suggesting 'transition-relevant places' at the end of a unit of meaning for the speech participants, indicating the transferral of turns and for the interpreter to take over. Although turn-taking is culturally conditioned and of course is a feature of the individual speaker's personality, it is nevertheless usually governed according to some kind of pattern which can be more – or less – orderly and allows for various degrees of speech overlapping (speaking at the same time). Turn-taking logic can be expressed highly schematically, as

- You said something now
- I want to say what I think about this
- I want to say it now:
  - I want to signal my agreement with what you said or I want to disagree with you
    OR
  - I will wait till you have spoken
  (adapted from Garzone and Rudvin 2003: 72)

The threshold for speech overlapping is highly culture- and situation-bound. Overlapping speech may be aggressive and intrusive (interrupting with one's own comments) but it can also be collaborative ('I agree with you and want you to know that while you are still speaking', realized by repeating the speaker's words, or phatic phrases like 'mmm', 'yes', 'OK', signalling agreement or interest through lexis and intonation). Another aspect of turn-taking that needs to be addressed is 'back-channelling behaviour' such as 'hm', 'really', 'well, well', or repair strategies such as 'I mean'. The interpreter must decide how to manage these verbal cues both verbally and pragmatically.

Even silence can be considered a 'turn', usually associated with a pause to reflect or with a polite form of non-acquiescence. Consider the following passage (*ibid*):

A. Would you like to go and ... uh ... get some coffee?
B. ...
A. Or aren't you in the mood?
B. What do you mean?

An interpreter must be trained to deal with the various manifestations of turn-taking and not least with the wide range of paralinguistic conversational features associated with these aspects of discourse management. A useful exercise is to have students talk together in loosely structured conversations that are as 'natural' as possible and to give one or two students in each group a task while another student translates. The task could be to interrupt, to ignore the interlocutor, to use long silences, to avoid eye-contact, to get upset/angry, to misunderstand, to use the third person, to ignore the client, and so on.

5.2.1   Identifying and translating units of meaning: mnemonic challenges

It is obvious that a translator/interpreter must avoid slavishly following the word order and structure of the source language, but it is not so easy to teach students how to identify the 'meaningful structures' in the source language utterance to be translated (a slippery notion but one which we might call, to make our point, the 'message'). Structurally, these *units of meaning* clearly do not necessarily coincide with that more artificial structural element – the sentence – but comprise words, parts of sentences, several sentences at a time, even a paragraph (see Strang 1968). Identifying units of meaning is done by scanning a speech, listening attentively to recognize and identify phrases or clusters of words that can be translated into corresponding clusters of words in the target language. The way words are organized into clusters in the source language will clearly not necessarily correspond to the way they should be organized in the target language.

Learning to identify meaningful chunks of speech – an utterance that contains a complete idea or notion and that is easily dividable into logical parts – is a skill that the students learn gradually. The trainer can vary both the lengths of the chunks as well as their division into logical parts to train this particular competence (see examples in the dialogues in chapter 6). In other words, when playing out a dialogue, or having the students translate from a text read by the trainer, she can read the texts in meaningful chunks to facilitate memory, or stop and start in a non-logical fashion to confuse the student.

One can upset the processing of units of meaning by adding numbers and names, thus upsetting the mnemonic retention of these chunks: students tend to hang on to names, numbers and lists and foreground these, sacrificing the primary focus of the utterance. This is of course because names, numbers and lists are hard to remember (precisely because they have no sequential logic or cohesive aids) and the student keeps these in her mind, forgetting the rest of the sentence. Another curious fact is that students tend to forget the last bit of a long chunk, perhaps because they are anxious to remember the beginning, or they are keen to start when they hear through intonation or paralinguistic features that the utterance is coming to an end. Memory retention, so crucial for reproducing these language factors, is of course very individual, but can quite easily be improved both in class and through exercises the students do at home.

5.2.2   Interrupting strategies

If the interpreter is not able to stop the speaker in time, she may have to ask for repetition of the last words. Interrupting and asking for repetition are two very important skills that the students should be learning through role play throughout the course. By talking for a long time, the trainer – or the other students – can deliberately provoke the student into interrupting. The students should be taught that they must never be afraid to ask. This also helps them become more aware of their own limits and work further on these. As mentioned, repetition should be requested through specific questions ('Could you please repeat *when* you placed the order?') rather than the bewildering non-specific form ('Could you repeat that, please?'). The students should be asking questions as clearly and specifically as possible. Sometimes they will be able to resort to gestures; one example of this was a case where a patient was helped to lie down on the 'stretcher'. The student did not remember the word 'stretcher', but was able to compensate for this temporary mnemonic loss by simply pointing to the (fictitious) stretcher. The ability to promptly understand whether the term they are not familiar with is absolutely essential or can be omitted is also a skill that students should gradually learn during the course. Role play in groups allows students more freedom to try out these techniques, while interpreting a dialogue acted out by the trainer in front of the classroom will probably make them more inhibited, but is also more realistic.

It is important to practise different kinds of interrupting strategies, ranging from discreet to intense eye-contact, a tactful cough, a touch on the arm or a gentle shift of the body towards the speaker, a movement of the hand, overt vocal interruption that can range from a soft 'mmm' to a firm 'Excuse me', etc., depending on the relationship between the speakers and the logistics of the room. If the class is ethnically and culturally mixed it is a good idea to exploit the different communication codes in role play.

5.2.3   Conversation topics and predictability

The flow of conversation is thus governed by the speech participants controlling and negotiating the conversation in various ways. One last aspect that the interpreter needs to be aware of is how topics are managed (certain topics that create cohesion through the text, i.e. topic management at the textual level) and which topics might require indirect strategies because they are considered to be culturally or socially inappropriate.

The listener's feedback is important in the development of conversation, and this also is to some degree predictable. She must develop an 'instinctive awareness' for how the form (turns) and content (topic) of a conversation evolve and to be able to predict to some extent what will happen next on the basis of the communication norms established for that particular situation and language. In this way she will be only half a step behind the speaker, as it were. Ersozlu (www.translationdirectory. com/article755.htm) suggests an exercise (devised for sight translation) that we find useful to help students practise 'predicting' roughly what they might expect from a given speaker in a given situation, which is also useful for terminology work. Give students the title of a text and ask them to tell the class/trainer what they expect from the text, using their passive knowledge of the subject (for example of the euro, she suggests). The students produce key-words by brainstorming, and are then given other key-words by the trainer and asked to make connections between them.

Thus topic management – and the breach of socially, situationally and culturally appropriate topic management – can be introduced into dialogues quite easily, as can adjacency-pair practice, not least in conversation openers and endings (greetings and farewells) and technical terminology.

5.2.4   Mode of address/direction: first person or third person?

Community interpreting and conference interpreting literature generally recommend using the first person when interpreting ('I have been sick for a week' rather than 'She says she has been sick for a week'). However, untrained natural interpreters and many service providers will spontane-ously use the third person to attribute speech to the speaker to whom the utterance/content 'belongs'. Students who have had no exposure to inter-preting tend to be a bit surprised by the fact that the literature generally recommends using the first person. First person 'quoting' prevents distor-tions from turning direct statement into reported speech by presenting an utterance in the same grammatical form as the original formulation. In some cases, depending on the use of pronouns and deixis in whatever language is being used, this may lead to a confusion of speaker roles, and difficulty in keeping track of referencing, of who is speaking and who is the author of the utterance, especially if the utterance is confused or dis-torted. In most countries, the use of the first person will be mandatory in courtroom interpreting. In 'real life' situations, when clients use the third person it is difficult for the interpreter to interrupt the session and request that the interlocutors switch to first person. In 'real life', there is a wide range of strategies used in actual face-to-face interpreting.

5.2.5   Instructions for the service provider: sharing responsibility for better quality performance

One interpreter we talked to gave us an excellent strategy for making the service providers aware of her role as interpreter and their own role as interlocutors. At the beginning of the session she brings a list of four or five 'rules' on a piece of paper that she starts to read out aloud – before anyone has a chance to stop her – that clarify her role in the communication situation: 'speak clearly', 'respect turns', 'don't use overly-technical language', 'the interpreter will interpret everything that is said by and to both parties' and 'the interpreter will speak in the first person' (personal communication; Petterson). This is very similar to what Tebble calls establishing the 'contract' (Tebble 2003). In busy institutional real-life settings, this can be a good way to compensate for lack of briefing. Such highly practical steps help make life easier for interpreters in their everyday work and are reassuring for students, who are still struggling with what may seem to be a daunting task.

5.3   Practical exercises: building competence and self-confidence

Although exercises should become progressively more difficult, at the beginning of the course it is a good idea to start gently, especially for those students with no prior exposure to interpreting.

5.3.1   Suggestions at the beginning of the course

There are a number of useful suggestions trainers can give to students at the beginning of the course so that they can make the most of the time available, also at home:

- Read extensively, both in L1 and L2, high-quality newspapers and news magazines and other well-written material that will help broaden your general knowledge.
- Strengthen your general knowledge of economics/business, the law, and medical-scientific concepts and principles.
- Keep abreast of the news through TV and radio in all working languages to be informed of general trends in both L1 and L2 communities: strikes and extreme weather changes, important political developments etc. These will often turn up in real-life situations, not least in the preliminary conversational phases of discourse in business meetings and 'polite chit chat', for example, and knowing what they are about will make them easier to interpret.
- Improve your public-speaking skills and practise making presentations in front of other people in both L1 and L2.
- Be prepared for life-long learning: be patient because bringing one's language and analytical skills up to the level required of a professional interpreter is slow work.

5.3.2   'Breaking the ice' exercises

Interpreting is a difficult skill in itself and performing in front of the class can be even more daunting. A good way to start gently is by using 'breaking the ice' exercises in class. These could be simple memory tests and sight translations in L1, starting out with general texts (for example from the day's newspaper), to listening exercises in L2 (internet, television, radio broadcasts) and then on to field-specific language and bi-directional memory and language production exercises. The duration and difficulty of each exercise should be gradually increased. Although it may seem obvious that technical texts are more challenging than everyday texts (e.g. from the newspaper), this is not always the case. Many students find the highly idiomatic, modern everyday language of general texts (for example in the daily press) more challenging than technical texts. This will depend on the students' language training and exposure to the L2.

5.3.3   Memory exercises

We try to incorporate memory games into the course, often at the beginning of the lesson to 'get the adrenalin flowing' and to get the students focussed. (Any number of simple memory games can be found on the web, ranging from visual memory to memorizing text to associative links and other strategies.) The first step to good recall is simply paying attention and concentrating, focussing on the salient information in the text, and this cannot be emphasized enough. Mastering cohesive memory strategies is also important: recognizing the links between sentences and units of meaning helps capture the flow of meaning by linking successive ideas. Thus concentration, comprehension, linking, recall and rendition are intrinsically linked. Simple memory tests are a good way to start getting the students used to listening carefully and memorizing: reading a simple text from the morning's newspaper in the students' mother tongue and then summarizing or paraphrasing the text in the same language (L1) is a simple exercise that can be used in any class situation. When they are ready, they can move from monolingual production to monolingual L2 processing (summarizing and paraphrasing in the foreign language) and finally to bilingual L1–L2 text production from their mother tongue to the foreign language. Again, we would recommend a progression from simpler to more complicated texts, shorter to longer chunks of speech or text.

A useful variation of this is summing up the whole text using only synonyms and paraphrasing (see below, section 5.3.10). The text could be about a topical issue, such as an actor's legal or medical problems, which is always entertaining and gives some light relief. Or it might be a current political issue with which the students are familiar; this always gets a good discussion going in class and is good for class rapport generally. Some students with no prior experience in interpreting find it very difficult to speak in front of the class and to interpret; at the beginning, working in small groups when doing these exercises, rather than in front of the whole class, reduces self-consciousness and builds self-confidence.

It also helps, we have found, to pay attention to the intonation and stress pattern while reading a simulated dialogue, as this helps the student understand better (i.e. reading aloud in chunks of meaning and logical units, rather than whole sentences). It might be useful – for pure memory practice – to use non-sequential text such as lists of names or cities. These memory exercises can be done jointly with the whole class (pick the 'volunteer' *after* you have read the text aloud so that they are all concentrating).

As with the dialogues, we found that students tend to remember the last sentence/clause of the utterance even in short texts and to forget the first part – due probably to the level of intense concentration. We have found that it is a good idea to practise identifying salient elements and to practise extrapolating sense from text, rather than semantic memorization. Finding logical links in-between the chunks of speech is important. When a student stops because she does not remember the following part, her classmates are asked to suggest a key-word that could help reconstruct the backbone of the speech or the relevant line of thought. This will help the students hear and understand the sense of the utterance and to reproduce this without being 'blinded' or 'hypnotized' by difficult lexical items. We would usually start with four or five lines and move on to seven, eight or nine lines towards the end of the course.

We also encourage the students to use memory games at home, especially while they are preparing for the exams. Memory exercises can be tiring and tedious, but it is important that the students realize how crucial this skill is (essential for accuracy) and that it is a skill that is not necessarily 'inborn', but can be improved through training. Doing memory exercises at home takes some of the stress out of the exam situation; stress is a major factor in exam underperformance for many students. Indeed, during exams, the students' greatest challenge seems to be not so much interpreting technique and terminology – they learn these quite quickly – but quite simply memory. This is why we keep stressing that memory exercises – either in class (these can be time-consuming), as homework or as exam practice – are so useful.

5.3.4 Chuchotage exercises

Chuchotage, which requires a great deal of concentration and very good comprehension skills, should be practised when the students have already worked on their memory and consecutive interpretation skills. One way to practise chuchotage is to ask the students to work in pairs, sitting quite far apart from the other pairs so that they do not disturb each other. One of the teachers then gives a brief ad hoc speech or lists the charges the defendant is accused of, for example, while one student from each pair whispers the translation into the other's ear. We then ask the student to repeat it in the target language while the others check if the whispered translation was correctly performed. The teachers can then comment on the rendition. Gradually students can begin to practise chuchotage in groups of three, where one student whispers into another student's ear and that student then repeats aloud to the third student what she has heard. An alternative exercise is to report or describe the utterance in the third person or to paraphrase it. This exercise can be done through L1–L1 or L2–L2 as a warm-up exercise to then proceed to L2–L1 and L1–L2. The text complexity should be kept to a minimum until the students feel confident enough to work with texts that they would find in real-life situations.

5.3.5 Enhancing listening and comprehension L2 skills at home: strengthening cultural competence and general knowledge as well as language skills

As mentioned earlier, listening to television or radio broadcasts and reformulating and/or summarizing the discourse in the students' own language, and later in L2, is a rewarding listening/comprehension exercise. For English L2 students, CNN and BBC World news broadcasts are useful for interpreting practice both at home and in the classroom because they also provide non-British, non-American English accents. The advantage of listening to international broadcasts such as programmes broadcast on BBC World and CNN is precisely the variety of international English, especially in documentaries from Africa and Asia. (Where other languages than English are involved students can watch or listen to national broadcasts in L2 if these are accessible.) Local broadcasts or television programmes that use field-specific and institution-specific terminology are also useful: there are television series broadcasting real trials, for example. It is in itself a useful language exercise for students to watch these programmes, paying attention to terminology (and in the court setting, to interpersonal pragmatics, participants' roles and institutional hierarchies). Monolingual interviews are good for bi-directional consecutive practice or even chuchotage. It is also good practice to get students to reformulate and/or summarize passages from trials or news broadcasts first in their own language and later in the L2. Interpreting spontaneous natural speech with all its inherent flaws can be extremely challenging, especially after the orderly, tame speech of the prepared dialogues.

There is also so much audio-visual material available now on the internet that it will probably become a primary source of material in the future. Trainers can guide the students towards useful websites, or they can find these websites on their own. They can access L2 websites for language training, broadcasts of real or fictional court- and police-related film material, patient–doctor consultations (fictional), and/or recordings of business meetings. Using online resources they can download programmes on their iPods and listen to them at leisure. These sources can be used both for comprehension practice and text-reproduction in various formats – consecutive, chuchotage, voice-over, and so on.

Using these resources, the students can stop the recording at any time, to listen again if there was something they didn't catch or a word they didn't understand, etc., and to improve their rendition. Practically any broadcast can be used for these exercises, but interviews are useful because they are so structured: they contain manageable turn-taking strategies (or sometimes challenging turn-taking techniques, for example in multi-party political interviews and talk shows) and they use a range of registers and discourse patterns. Furthermore, interviews, as mentioned, may contain a variety of emotional reactions: surprise, anger, irony, respect or courtesy, which are all governed by specific discourse strategies. Interviews with television or film personalities may contain culture-specific humour which can be very difficult indeed to translate and therefore provide an interesting exercise. These exercises are also useful to do at home as memory practice before exams. This listening practice can be further developed by having the students bring back reports and summaries to class.

5.3.6   Keeping abreast of new features in language and society

Listening to programmes such as these will also expose the students to neologisms – especially to technological neologisms and technological 'slang' – and technological developments generally. These may be pertinent to their own future academic and professional development (e.g. developments in and neologisms describing the computer and internet world, sophisticated research tools, terminology searches and specific online resources). As well as neologisms, students are thus exposed to norms of political correctness in the L2, which may be valuable for their language production skills as interpreters and which may be very different from their own country or culture. It also gives the students a better idea of what issues people in the L2 culture are interested in. Lastly, it boosts general cultural awareness and builds cultural competence: news broadcasts from all over the world also give valuable information about different customs and attitudes, and are often followed by in-depth reports and documentaries.

5.3.7   Listening practice: the importance of L2 comprehension

We feel that the difficulty and importance of L2 comprehension is often underestimated. One of the basic assumptions in mainstream interpreting studies has been that the interpreters should interpret into their own language, the assumption being that the delivery is easier, more accurate and more fluent than interpreting into the foreign language. In our experience, however, we have found that this does not always hold true. The comprehension of the L2 (even – indeed especially – in current everyday language) is often more challenging than *reformulating* in L2. This stands to reason: if the student 'perfectly' understands the source text it is practically always possible to reformulate it in L2 even if the resulting L2 target text is simpler than the original, or even clumsier. It is still accurate. Thus, rendering the message from L1 to L2 is not as challenging as it sounds, even if the students are not used to speaking in L2. If, however, there are terms in the L2 text that the student does not understand, she will simply not be able to reformulate it in L1, however perfect her command of L1 is. This pertains especially to dialogue/face-to-face interpreting. In other types of discourse/interpreting where the quality of the rendition is crucial (in persuasive texts or texts where form is as important or more important than content, such as political speeches) this claim would have to be qualified.

5.3.8 Sight translation and reformulating

Many interpreting trainers use sight translation in class to familiarize the students, especially those with no prior experience in interpreting or translation, with the language-transfer process and to get them used to working quickly and making on-the-spot translation decisions. It is also a gentle way of easing them into performing in front of the class (i.e. a 'soft' public speaking-cum-interpreting exercise). Furthermore, it may train students to anticipate translation (terminological, syntactic or stylistic) problems by developing reading techniques whereby the reader's eye is slightly ahead of their voiced rendition, and thus to anticipate the decisions they will need to take a few seconds later. Sight translation is used in many areas of public service and dialogue interpreting and is thus useful both as a professional skill (legal texts, for example, can be very challenging) as well as a language exercise. In sight translation, reading rather than listening skills are developed, and rapid-reading practice – to scan a text and grasp the gist of it – is easy to practise in class and at home.

All kinds of texts can be used for this exercise – even the morning's newspaper. A 5-minute sight translation at the beginning of the lesson gets the 'adrenalin flowing' and gets the students focussed on their tasks. Later on in the course it is a good idea to use texts that correspond roughly to whatever the theme of that part of the course happens to be.

Students who have already attended interpreting courses may not find sight translation so challenging, but those who are doing it for the first time may be very slow, translating one sentence at a time and then stopping when faced with the first term with which they are not familiar. As mentioned earlier, students must learn to identify those elements that are essential to the overall meaning of the sentence or paragraph. Analysing the text and breaking down messages into units of meaning, and then reformulating them is a good exercise. At the end of a sight translation exercise, students can repeat the exercise by using only synonyms or different expressions to convey the same meaning. They can start from the end of a paragraph and sight-translate 'backward', that is by starting not from the last word, but from the last concept expressed at the end of the paragraph. This is a sort of obstacle race that will train them to play with words with a purpose in mind: i.e. never translate word for word or *mot à mot*, but stick to the core message. At the end of the sight translation exercise one of the students can be asked to sum the whole text up.

As mentioned on p. 114, we found some useful exercises by E. Ersozlu on http://www.translationdirectory.com/article755.htm, for example the following reading exercise. The students are asked questions about a short text they have scanned – general questions followed by specific questions – and then re-read the text more quickly and repeat the exercise. The third time they read the text carefully and are asked more in-depth comprehension questions. The following site by the National Council on Interpreting in Heath Care (NCIHC) also contains useful information on sight translation in the health care sector: http://reachnola. org/pdfs/Sight%20Translation%20and%20Written%20Translation.pdf

5.3.9   Vocabulary practice

Useful vocabulary-enhancing strategies for students at the beginning of the course that can be associated with interpreting or sight translation practice might be to:

- Decide if the word is really needed to understand the text (i.e. is it a key word?) If it is not, move on. If it is an adjective or an adverb you can easily construct a meaningful sentence without it.
- Decide if the word is repeated in the text several times either in the same form or as synonyms; if so it is probably a key to the global message.
- If a word is unfamiliar, use the context surrounding the word to guess its general meaning, or analyse the parts of the word to guess its probable meaning.

- Remember to focus on nouns and verbs which are generally more important to the basic understanding of a passage. A high degree of nominalization and abstraction, more typical of some languages and some genres than others, can make a text very heavy and difficult to understand and process, and being able to identify the core constituents is useful.
- Become familiar with adjacency pairs in the L2 (where one utterance has a role in determining the next one, as in question-answer) is useful because it greatly eases the language flow and interpreting speed; being able to predict adjacency and collocation and to reproduce them instantaneously and automatically without too much deliberation considerably helps speed and fluency.

These exercises do not necessarily train semantic translational accuracy, but are good practice for making quick decisions and moving forward when students get stuck on words they do not know, and they train the ability to predict more accurately what is to follow.

5.3.10   Paraphrasing and summarizing

By 'paraphrasing' we mean the restatement of an idea which uses different vocabulary and a different sentence structure from the original. When a cultural reference or an allusion does not make sense when it is translated literally into the target language, or when an interpreter cannot think of the exact expression in the target language because she is under time-pressure, then a paraphrase, or an approximation, can be used. Paraphrasing is an important skill for translators and interpreters because it helps them to reformulate terms for stylistic reasons or, more importantly, when the languages do not possess corresponding terms.

   Another function of paraphrasing is to provide alternatives to words that are culturally taboo, or socially unacceptable in certain cultures (i.e. the use of euphemisms). This is particularly pertinent to the medical sector, where students can perform a dialogue that revolves around a medical examination with a Western doctor and a patient whose culture does not permit her to verbalize descriptions of parts of the body too directly. But euphemisms and paraphrasing might be necessary in less obvious cases too: the expression of pain (both the degree of intensity and the verbalization of pain), and the description of symptoms, are profoundly culture-bound and may need a good deal of adjustment through paraphrasing (see Galanti 2002 for excellent examples of how pain and other health-related features are described differently by different ethnic groups in American hospitals). These are excellent springboards from which to discuss cross-cultural aspects and the inextricable bond between language and culture. Summarizing too is an important pedagogical tool that helps students learn to grasp core meanings quickly. Useful exercises are listening to speeches and orally summarizing the main points, writing summaries of news articles, deciphering and summarizing difficult texts, and explaining complicated concepts in a way that is comprehensible.

5.3.11 Terminology

At the end of each session of training within a 'setting', whether business, health, or other, the students should have a list of the relevant expressions to keep as a resource. Ideally this should not be simply a list of words – as this would be difficult to remember – but rather a set of *expressions*. (See the bibliography for useful terminology websites.) In the medical setting, for instance, the list includes all the different ways to define a 'disease', or a 'complaint', the main medical professions, different types of medical tests, as well as expressions such as 'to have a test done' (patient), 'to run a test' (physician). In the business setting, if students hear the word 'contract', they should automatically think of 'drawing up a contract' or 'signing a contract', or if the word 'meeting' is uttered, they immediately know that a meeting can be 'planned', 'convened', 'scheduled', 'postponed', etc. Another useful terminological and stylistic skill is learning and remembering synonyms, i.e. to have a whole set of synonyms ready for each setting, a ready-made set of solutions at hand.

5.3.12 Practising register variety: verbal and non-verbal aspects

Register variety can be related to interpersonal relations in institutional or commercial settings and the power asymmetry involved in these relations, as discussed in chapter 2. These can be examined with students in a cross-cultural perspective using examples from their own countries and the other countries they are familiar with. Students should be exposed to a wide variety of registers that occur naturally in interpreting settings and then reproduce them in role play.

We normally start with very simple examples, like the following: If a school teacher is bothered by a too noisy class, she might say any of the following: 'I would be grateful if you would make less noise', 'Please be quiet', or 'Shut up!' Then we pass on to examples taken from texts or dialogues used in class. In the legal setting, for instance, in Dialogue 12 in chapter 6 the interpreter who is interpreting the magistrate's statements to the two foreign defendants whose English is quite basic will immediately perceive the differences in register. The Italian magistrate will no doubt use a very formal and high register in Italian, while the two foreigners – as we see from the statements reported in the simulated dialogue – use very colloquial terms. This normally leads to a discussion in class on the need – or not – for the interpreter to change register in order to be understood by the interlocutors. It is a simple but good example of register variation, alternating between talking to the defendants and maintaining a high register, thus respecting the norms of formality appropriate to this setting.

5.3.13   Public speaking

Public-speaking skills are important for student interpreters simply because they need to practise performing in front of other people. They must also learn to be assertive, self-confident and credible to ensure that their interlocutors trust them (and thus trust their interpreting rendition). Interpersonal non-verbal features such as facial expressions, posture or eye-contact are vital cues for any speaker who has to build a trust-based relationship with his/her interlocutors.

General 'golden rules' of public speaking include features such as being engaged and, preferably, friendly, not reading slavishly from the script, using examples and anecdotes, keeping eye-contact with one's audience, being prepared for the assignment, and using and responding to body language. These are all good rules of thumb that are useful when speaking in front of an audience, but not all of them are immediately pertinent to interpreting. However, skills such as not reading from a script are useful for interpreters too, because it trains them to engage with their interlocutors (in terms of body language and eye-contact) rather than being exclusively focussed on the text (be it spoken or written), and also not to be too dependent on notes.

Of course, in the type of settings we are dealing with in this book it is important to remember that features such as body language and eye-contact, especially in an intimate setting with three or four people (rather than a large audience) should be adapted to the culturally appropriate norm in that particular cultural and ethnic group(s).

Other public speaking 'rules' that improve interpreting delivery are clarity and succinctness, i.e. to be clear, brief and to the point. Although an interpreter must always follow the source text or utterance, of course, learning to be clear and brief is a good way to practise grasping the core message and reducing obvious superfluities (which is a difficult enough evaluation for an interpreter to make), as we have also suggested above.

Controlling the quality of voice and avoiding distracting fillers are also directly pertinent to interpreting delivery. Also, just as in public speaking and indeed in any speech, students should be reminded that simply by changing the rhythm, volume and tone of voice one can stress key words and concepts.

A useful exercise, though clearly only possible in small classes, is for students to prepare 10–15 minute presentations on a set theme (adjusted to their L2 level: e.g. an industrial product presented at a trade-fair, their home town, a product or event typical of their home town; or for a more advanced class, presentations about political or financial themes) preferably in PowerPoint format. At the beginning of the course, we ask the students to start preparing their presentations and bring them on a pen-drive to class, so that any extra class time can be used for presentations. They are given a few suggestions on what to present and a few suggestions on how to present and refer to their slides, such as 'in the top left hand corner' and also more generally 'Today I'd like to tell you about', 'On this slide you can see'. This allows the students to practise their language skills, presentation and text-organizing skills, field-specific terminology, and a variety of registers and genres. It is also extremely useful in fostering self-confidence. We do not use student interpreters for these sessions, because it disturbs the presenter, but we do use the presentations for memory exercises by encouraging the students to ask questions afterwards and then to summarize what their colleague has said in L2: thus, the students are more concentrated on the presentation and are focussing on trying to remember what is being said. Obviously, the students also get feedback on their language errors.

Another variation is to form small groups where each student tells the others about a fact or event from recent news. They decide together on a topic and agree on how to present it in the other language, then they form new groups (different combinations) and each student relates the same events in L2.

Preparing a student debate on a controversial issue (euthanasia, national politics, immigration, the role of women, the environment, whale hunting, nuclear power) is also a useful public speaking exercise. This trains the students to present delicate information and arguments, to defend and to conclude by balancing various opinions against each other. Another useful exercise is to have the students listen to a formal speech in L2 on a given topic and have them choose 10–15 typical discourse markers and 15–20 field-specific terms to use in their own presentations.

5.3.14   Glossaries

While preparing their dialogues for group work, students should be preparing glossaries at home (simple Word or Excel format is perfectly adequate). We do not usually assess students' glossaries, considering them as 'work in progress' to be continually updated.

5.3.15   Note-taking

This chapter will not address note-taking in any depth. We refer the reader to, for example, Gillies (2005). A simple and user-friendly note-taking manual specifically for interpreters in the public sector is *Note-taking for Public-Service Interpreters* by Heimerl-Moggan and John (2007). Taking notes is in itself a mnemonic aid, of course, and thus very useful as a memory exercise, so incorporating it even cursorily into the course can be very useful. Trainers who are themselves trained in or have experience with note-taking (especially those with a background in conference interpreting) will be in a position to teach this skill properly from the start of the course, but if pressed for time or lacking formal training in note-taking techniques, they can nevertheless offer general recommendations to the students that will help them find their own personalized note-taking system. Improvised notes can sometimes turn out to be misleading, so they should be used with caution. Taking too many or too full notes and concentrating on processing these visually rather than the rendition is counterproductive. Rather than writing full words, professional interpreters use signs, or symbols, to denote words and concepts. These symbols are often 'personalized' so that the interpreter recognizes them at first glance, but they are often only fully meaningful to that person. Clear simple notes/symbols and good comprehension will make the rendition more immediate and fluent and there will be less need to rely on the visual support of notes.

The most important things to remember here are that for the students' notes to be accurate and reliable, the symbols/notes must be written:

- in such a fashion that the student can immediately recognize them and understand their meaning;
- in a form to aid quick recall and thus to reproduce in real time.

Students should also be encouraged to jot down numbers, dates and names. (Remind the students to ask clients for a list of names, if possible, before a 'real-life assignment'. Foreign names can be very difficult to capture and remember.) The trainer can progressively increase the number of mnemonically challenging items during the course of the semester and then test the students to see if they are able to handle these without notes. During the retrieval phase, the students should learn to only glance down quickly, and start speaking when they look up again and then glance down again for visual aid, especially of numbers, figures, names and places, only when necessary. Practising reading ahead, as in sight translation, is a useful exercise. Another useful exercise is to work in pairs and have one student read a short paragraph from a reasonably simple L1 text and the other student translate consecutively, the first time taking copious notes, the second no notes, the third only numbers and dates, etc. More exercises can be found in Heimerl-Moggan and John (2007).

5.3.16   Cross-cultural simulation games

Cross-cultural simulation games are an excellent exercise to raise awareness about how people from different cultures and communication systems have trouble understanding each other and how this can lead to serious cross-cultural misunderstandings despite the good faith of all the interlocutors involved.

The advantage of these games is that rather than just discussing cross-cultural differences, they enable the students to actually 'feel' how frustrating this lack of communication is. One of the simulation games we have successfully used in class is *Barnga: A Simulation Game on Cultural Clashes* (Thiagarajan and Thiagarajan 2006). The surprise twist at the end of the game (which we will not reveal here) is an excellent platform from which to discuss how people speaking different languages and embedded in different cultural and communicative systems simply are not able to reach an understanding.

The aim of this game is thus to teach the student not just how complex intercultural communication is, but how confusing it is and how vulnerable it makes the parties. It also demonstrates how the rules of communication in each cultural group are different and how this can lead not just to misunderstandings and extreme confusion, but ultimately also to tension between the parties. It is thus a perfect metaphor for the degeneration of simple cultural and language-based misunderstandings into ethnic conflict. Although it is revealing of the 'darker' sides of cross-cultural communication, if used well, with a good discussion at the end of the session, it can be highly constructive and highly motivating in that it can be used to train students that cross-cultural non-alignment and misunderstandings can be solved, challenging them to find effective solutions in real time to the problem of communicative misunderstanding.

Barnga requires quite a bit of practical organization, but is worth the effort and can comfortably be fitted into a single long lesson (minimum one-and-a half hours). Ideally, it should be played with no more than 20 students, but we have managed, albeit with some difficulty, also with very large groups (60). It requires a large room for the many tables and chairs that are needed for the class to be divided up into groups of five or six participants, and as many sets of ordinary playing cards as there are groups of students. We will not reveal how the game is played, but instructions can be found on http://socrates.acadiau.ca/courses/educ/reid/games/Game_descriptions/Barnga1.htm or in the original booklet. Trainers should include a period of de-briefing after the game, because some students may experience a small 'culture-shock' and they need time to digest this emotionally and cognitively. It also gives the trainer the opportunity to apply the game to real-life situations and to ask questions like: 'How does not being able to speak the same language contribute to one's feeling of vulnerability and alienation and lead to misunderstandings?' and 'What real-life situations and problems does this game remind you of and how would you solve them?'

There are other simpler games that can be practised as 'warm-up' exercises at the beginning of class to help students focus and concentrate and that reveal cross-cultural differences in body language, the use of personal space, eye-contact, voice, and so on. (Interactional role plays, cultural contrast, and simulation games can be found at: wilderdom.com/games/MulticulturalExperientialActivities.html.)

5.3.17 Using training videos in the class

There are a number of training DVDs on the market. We have chosen two videos that we have used year after year because they illustrate so well, in our opinion, vital issues of interpreting. These are *Mental Health Interpreting: A Mentored Curriculum*, by Robert Pollard (1997–1998) and the dated, but still very popular, *Points of Departure. Ethical Challenges for Court and Community Interpreters* (Vancouver Community College 2000), which provide useful illustrations and discussions on ethical as well as practical issues – such as overlapping speech, interrupting, asking for repetition, assertiveness and professional role. The Australian interpreting association (Ausit) has also published a series of DVDs that are helpful as training material: *Unethical? Who me?*, *Dealing with Stress while Interpreting* and *Interpreting and Translating for the Police*. In addition to raising a number of essential issues regarding professional practice and the code of ethics, it is good for the students to be able to visualize exactly how interpreting is performed in practice: in the doctor's office, in the courtroom, at the police station, and so on. See the bibliography for a list of audio-visual materials that can be used in the classroom.

5.3.18   Useful material: research tools and sources

Original documents such as police and medical reports can be used very effectively for sight translation exercises as they provide real-life situations, with valuable information not only on terminology but on institutional procedures and internal organization. The documentation can then be used to write dialogues. When using original documentation of any kind all the names and (personal) data should of course be changed to maintain confidentiality. For the legal setting we often use material based on high-profile cases like the O. J. Simpson case (although this is now dated), as transcripts of original material from cases like these can easily be found on the internet, as can transcripts of excerpted jury trial testimonies from countries that allow the publication of such data (see www.courttv.com, www.courttvcanada.ca, http://news.findlaw.com and http://www.un.org/icty). There is an endless amount of material for sight translation and memory exercises that can be found on the internet, but we also use leaflets and advertising material from trade-fairs, meetings, shops and demonstrations that we have actually been to and can thus contextualize in a much more hands-on manner. For medical terminology, both trainers and students can find excellent training material in brochures at the chemist, local health care Trusts, doctors' offices, patient-based associations (blood donors, cancer survivors, hospital groups, patient support groups, transplant recipients' groups, rape and abuse victim associations, crime prevention centres, and so on).

One of the trainers participated as a language expert in a Leonardo da Vinci european project called *Medics on the Move* (MoM). The MoM website (www.medicsmove.eu) contains more than 1,000 everyday medical terms in six target languages (English, Danish, Dutch, German, Italian and Swedish) and searchable databases of more than 200 workplace-oriented communication scenarios in a cross-cultural medical context. It provides medical professionals using L2 and L3 with communication tools and online advice to help them function effectively as professionals. A Swedish doctor or nurse who is practising in the UK, for example, has problems understanding a patient and consults the MoM site. MoM offers insights into medical communication and socio-cultural interaction, in addition to providing language tips, links to work rules and regulations, and medical guidelines. The MoM programme includes doctor–patient interactions during different phases of consultation, and consultation with colleagues in the course of the working day, and more. Although it is an online tool meant to help doctors and paramedical staff working in foreign countries to improve their linguistic skills in the local language, it is valuable material for both trainers and students working with cross-cultural and multilingual communication in the medical sector (the lesson will obviously require online access for all students). One particularly useful tool is a wordlist and a syllabus in the form of a long list of dialogues: doctor–patient, doctor–doctor, doctor–other healthcare worker. These dialogues touch upon several aspects that are potentially useful for interpreter training, such as patient history, pain, surgery, referral to specialists, etc. Participants can also add new words or synonyms and record their pronunciation if the problem is due to lack of understanding of a particular term. Once the students have learned how to enter the programme they can then use this tool alone. One group of students even set up workshop forums online where they could exchange their opinions and make their contributions.

5.4 A guide

We have included here a very practical, prescriptive, student-friendly guide to interpreter performance in a somewhat humorous format that we hope can help students increase their self-confidence and assertiveness. It is based on our experience as interpreters as well as trainers and makes points that may seem obvious or prosaic, but that are fundamental and about which students require constant reminding and reassurance. This 'guide' is best used at the end of the semester when the students are familiar with theoretical issues and have had sufficient interpreting practice.

The Hitchhiker's Guide to Effective, Safe and Ethical Interpreting: Eight Basic Rules

1. BE PREPARED. Get as much information as you can before the session; if you need to, brush up on terminology (depending on the situation, there are many sources: glossaries, newspapers, maps, the internet, brochures for public services); you may have

to insist because service providers are often reluctant to give information, not only because the information might be of a confidential nature, but because they don't have time or they simply think it is unnecessary.

2. INTRODUCE YOURSELF.
3. LISTEN CAREFULLY. Always listen carefully to the speakers.
4. TRANSLATE AS ACCURATELY AS POSSIBLE. Translate as accurately as possible, but not necessarily literally. Follow the message rather than the words and follow your common sense: you're the language and culture expert. Aim for a global message transfer rather than concentrating on individual lexical items. Follow your instinct, remember that your 'instinct' has been trained to deal with this kind of neurological processing – trust it, and concentrate!
5. ASK. Ask if in doubt – ask and lose face rather than lose accuracy.
6. NO TALKING TO THE CLIENTS. Avoid private side-conversations unless strictly necessary (such as asking for clarification, or establishing that you speak the same dialect/language, before the session starts), and in those cases explain to the other party what you're doing. Avoid being alone with a client, it can get you into trouble.
7. FOCUS. Concentrate 100% for as long as you're able, and don't get distracted; this makes all the difference for text retention and memory. If you're losing concentration, speak more slowly to slow down the communication until you get your breath/focus/ concentration back, or take a few seconds' pause after the last change of turns. To stop the interlocutors from speaking during this phase make it clear that you're concentrating – perhaps by looking at your notebook or even closing your eyes – to avoid communication for a few seconds so that you can get back on track. If you lose concentration in mid-sentence, then stop at a difficult lexical item as if you're searching for the equivalent just for three or four seconds till you get your concentration back. This will also slow down the communication event (which is a good thing and might get them to speak more slowly). It will also help draw attention to you, not as a primary interlocutor, but as the person whose responsibility it is to transfer the message and will remind them how difficult and important this process is. It may sounds obvious, but it isn't: stay focussed on the theme of the conversation, this will help you contextualize what you're hearing.

8. BE READY TO IMPROVISE. Sometimes unpredictable things happen: the rapport between the interlocutors changes suddenly, completely unfamiliar concepts or lexical items crop up, or other unexpected events. Try to use your knowledge and experience creatively to find a solution; improvise and paraphrase, if you need to, but don't lose accuracy and don't lose control of the communication. Remember that the different interlocutors' intentions, expectations, goals and understanding of the situation may be completely different from each other and this might cause confusion. Don't let it throw you!

5.5   Assessment

As mentioned, there are a number of recent studies on the testing and assessment of interpreters that address these issues far more rigorously and thoroughly than we are able to do in this volume. Angelelli and Jacobson's (2009) edited volume *Testing and Assessment in Translation and Interpreting Studies* (which includes sign language and conference interpreting) addresses a number of factors, beyond simple terminological accuracy, that should be included in any global evaluation of interpreter performance – such as measuring interactional competence and cohesion, register and interpersonal aspects. The book takes a highly rigorous scientific approach based on empirical data and sophisticated testing models, and thus aims to go beyond more intuitive and often vague assessment criteria. It also includes a number of essays on various assessment instruments.

Another study that addresses conference interpreting performance specifically, in a scientifically rigorous manner, is David Sawyer's (2004) volume *Fundamental Aspects of Interpreter Education,* which contains several chapters on various aspects of interpreter assessment. Such works are no doubt potentially extremely useful to community interpreter trainers, but many trainers in our field today do not have the background in conference interpreting that an application of these models and criteria demands. As the discipline of community and/or dialogue interpreting develops over the next decade and beyond, it too will hopefully be able to provide equally rigorous scientific assessment tools.

Because the present volume is of a highly practical nature, we will simply limit ourselves to mentioning some of the most basic methods and criteria that we have adopted in the classroom. Our assessment methods are simple, consisting of a theoretical part and a practical part, as detailed in the next sections.

5.5.1 Theoretical component

We test the students' theoretical skills in one or several term paper(s) or essay(s) handed in at the end of the module. We allow for a certain amount of flexibility, allowing them to choose a topic themselves, or a particular problem or angle that has caught their interest, as long as it is pertinent to the course contents. This freedom allows the student to do a number of things that a regular written exam or oral test would not: to identify and reflect on what most interested them during the course; to develop these themes through bibliographical research at home (libraries and internet); to discuss theoretical issues with fellow students; or to gain a better knowledge of their local context through practical empirical studies and interpreter questionnaires at the local hospital, courts, businesses, etc., thus developing skills necessary for qualitative and quantitative analysis and using and developing their imagination and creative skills. Essay writing gives them the freedom and time to develop all these skills in the peace and privacy of the working environment of their choosing (library, home, medical or legal institutions for empirical work, etc.).

Almost without exception we found that students take to this assignment with great enthusiasm. Interpreter ethics turned out to be one of the most popular topics. Another possibility is to write a short essay on interpreting skills, for example 'How my memory works and how I can improve it'. It is also invaluable material for us as trainers because it gives us information about what the students find most interesting or challenging and where their strengths and weaknesses lie. We usually provide the students with a list of suggested topics, guidelines on how to write the essay and the bibliography, and a warning about copying from the internet.[4]

Students' theoretical skills are further assessed through an oral exam testing their knowledge of set textbooks, for example Gentile, Ozolins and Vasilakakos (1996) and Hale (2007). Unfortunately, time constraints often preclude in-depth discussions with the students about theoretical and ethical issues discussed in the books, but in our experience, if they are given the opportunity, the students enjoy talking about and demonstrating what they have learned during the course, especially if they are asked directly about what most interested them, or are themselves encouraged to ask questions. If it is pertinent and appropriate to the global objectives of the course, the trainer can also evaluate the students' delivery skills and L2 proficiency, rather than just the content.

The ideal situation would be to have a continuous assessment situation whereby the students are graded during the course not only on their performance in class (sight translations, memory and text production exercises, public speaking and presentations, etc.) but also on tasks they are given to complete at home or with fellow students and then record in audio format. The ideal combination would thus be to have a combined portfolio of essays and oral production. This allows the students to analyse their own knowledge self-critically (having received feedback from trainers on their essays) and to have an objective towards which they can work (the end-of-year essay) on the basis of their own error-analysis and the trainers' feedback (see Kainz, Prunc and Schögler 2010, for assessment through portfolios, and see Slatyer 2010 for self- and peer-critique in class). Portfolios allow the trainers to assess progression throughout the year in terms of research skills and increasing maturity as researchers as well as interpreters. As well as essays, students can also compile glossaries in one or all of the fields that have been covered. The feasibility of this method is again dependent on time constraints and class size.

5.5.2 Practical component

Our preferred exam format consists of three components: an oral discussion of theoretical and professional aspects from the textbook(s), the interpretation of a brief dialogue from and into L1 and L2, and memory, text-production (summary, paraphrase in L2) and/or sight translation exercises. With large student numbers, to avoid writing and using many dialogue versions, it might be an idea to keep the students who have already finished their simulations in the room so that they do not discuss the dialogue with their fellow-students who are waiting to take the test.

Because we aim to assess the students' skills in a hypothetical workplace situation, we need to evaluate not just translational accuracy (in a very strict sense of the word, without entering into the timeless and irresolvable problem of faithfulness and of semantic versus communicative accuracy), but also the students' overall performance and ability to implement all the linguistic, cultural and interactional skills they have learnt during the course. For a trainer assessing a student's performance, it is very difficult indeed to evaluate all of these factors, not to mention finding objective evaluation criteria. Cultural competence is particularly difficult to evaluate, but potential misunderstandings can be introduced into the dialogues to test the students' ability to find problem-solving strategies under stress. The following (Table 5.3) are a few simple suggestions on possible assessment criteria with hypothetical trainer comments. In the table, 1–5 indicates a possible rating system, 5 being the highest score.

Table 5.3 General assessment criteria

Rendition criteria	L1 mark (1–5)	L2 mark (1–5)	Comment	Examples from dialogues
Comprehension (especially of L2)				
Translation strategies; Terminological accuracy (errors and mistranslations, omissions, additions)			Tendency to omit what seem to be redundant elements, due to lack of comprehension, but also some key passages/tends to elaborate on the text to hide confusion/ misunderstanding.	
Language proficiency				
Articulation/pronunciation (aspects like voice projection could also be assessed here – mumbling can be very difficult for the interlocutors to process)			Does not articulate well, the listener has to concentrate hard to follow.	
Speed			Speaks too quickly, hard to understand/ hesitates too much – listener loses track of dialogue.	
Fluency in L1 and L2			Native speaker, but speaks very hesitantly under stress. Un-idiomatic language.	
Text production skills				
Cohesion and logical sequencing			Doesn't use enough links between sentences – hard to follow the logic/argument.	

Register Contextually appropriate language (for example, politeness strategies)	Register too low for the type of text Doesn't use enough politeness strategies – source language speaker comes across as rude/aggressive.
Cultural competence Comprehension of culture – specific elements in L2	Misses some of the more nuanced communication features such as irony and humour.
Production of culture-specific elements in both L1 and L2	Translates culture-specific elements such as humour too literally – the text sounds very stilted.
Technical interpreting skills Memory	Tends to forget and ask for repetitions after short chunks.
Interrupting strategies	Never asks for repetition, but omits too much/too aggressive in asking for repetition.
Delivery, personal attributes (voice, body language, self-confidence)	Nervous and hesitant – affects performance and might jeopardize trust placed in interpreter by interlocutors.

If it is not possible, for practical reasons, to record the exam session, we have found it useful to make a photocopy of the dialogue for each student and then write comments on it as a mnemonic aide for the trainers, who will give it back to the students as feedback after the test. It is also helpful for the students to know what the criteria are before the test. This is not only useful in making the trainer's own assessment more reliable but very useful for the students themselves to enable them to understand their own weak points – which may either be the result of exam nerves or more structural weaknesses.

To what degree should L2 proficiency, as well as or rather than interpreting skills (if indeed these two can be separated), be assessed? What the trainer needs to look for, and what we have attempted to suggest in the table, is the interface between the two – that is, where language comprehension and proficiency in both L1 and L2 do not just affect, but constitute, interpreting performance.

5.6   Students in the workplace

5.6.1   Placements

Ideally, students should be offered a placement period at the end of their course, depending of course on the training institution's agreements with private and public institutions. Appropriate placement settings would be the Chamber of Commerce, private firms and corporate export offices, tourist offices, hospitals and police stations. The opportunity to work alongside official interpreters is of course the perfect learning opportunity. Our experience with placements at hospitals (the emergency room, gynaecology and cardiology) and local police headquarters has been extremely positive, giving the students the opportunity to assist professionals and also to try their hand at interpreting themselves.

5.6.2   Making connections between the classroom and the community

We have found that students are more motivated when they feel that there is a tangible connection between what they are learning in class, the local community, and their own futures. Those students who do practical research papers on interpreters working in their local areas (hospitals, firms, police, etc.) and interview them directly already have quite a good idea of how language services function in institutions and in the workplace, and are encouraged to share this with their classmates. This is a good way of fostering a positive bond between interpreter training institutions and the local community.

In many faculties, some courses are highly theoretical and do not necessarily address concrete professional issues and we feel particularly privileged to be able to teach a skill that is both theoretical and practical and that the students believe will have some use for them in building their futures. Not least, many students feel gratified to know that this is a skill in which they can in some way serve their local community. Indeed, issues that touch upon immigration policies and multiculturalism appeal particularly to students because it makes them feel that they are part of a global community where they can make a contribution to foster integration through their skills as communication facilitators between cultures and languages.

6
Annotated Dialogues

In this chapter the reader will find fifteen annotated dialogues, five for each sector (business, medical and legal settings), which are designed to familiarize the trainer with the use of scripted simulated dialogues in the classroom. They can be used as a guide to creating similar dialogues, they can be used as they stand, or they can be adapted to suit the trainer's specific requirements.[1] The points for comment are numbered in bold in square brackets in the text and the comments themselves are provided at the end of each dialogue. The dialogues are in English and Italian, but a translation of the Italian text has been provided so that they can easily be adapted to other language combinations. Rather than using near-literal and awkward-sounding gloss translations, we have attempted to provide real translations that read well and sound reasonably natural in the English so that they can be used as natural-sounding source texts for yet another language pair. Needless to say, authentic spoken dialogues are the optimal teaching instrument, but these may be difficult to come by, especially in the necessary language combination. Some of the dialogues in this chapter are based on real events in which the trainers were involved as interpreters. Scripted dialogues, although they are less natural and more stilted, allow the trainers to include precisely those elements that their students need to work on, be they terminological, conversational or interpersonal aspects. Scripted dialogues may not be verbatim accounts of the episodes that have actually taken place, but they do reflect trainers' aggregate experience as professionals and as teachers, and we believe they are therefore a useful methodological tool.

The Italian–English language combination restricts our scenario in terms of 'community' interpreting, since standard UK, US or Australian English is not the typical community interpreting language in Italy.

We could have used Filipino or Nigerian English in the health and legal settings, or Japanese speakers of English in the business setting, but it is not always easy to reproduce non-standard English effectively in writing. Trainers can vary these scripts by adding other varieties of English or English used by non-native speakers.

These dialogues are also meant to function as points of departure for discussions on language- and communication-related features such as:

- redundancies – if and when omissions are permitted
- explication/ clarification – if and when additions are permitted
- language-specific changes in lexis and syntax
- floor-management and speaker-coordination
- institutional register and informal register – colloquial versus formal language through lexis but also through modality, indirectness, hedging (also switching between the two: negotiating social roles)
- politeness markers
- phatic speech and gap-fillers
- body language
- discourse features such as repair, back-channelling, adjacency pairs etc.
- culturally appropriate language – cultural and social taboos
- ethical issues.

The dialogues are also useful for looking at how the power relations between the interlocutors are constructed (who is in charge and how this is signalled linguistically – who asks the questions or decides on topic transition, and so on). To help the teacher use these dialogues more efficiently in class, we have provided an introductory note on each of the dialogues regarding the level of difficulty for general language and technical terminology – and thus whether they are more suitable for the first lessons of a course or the end of the course. As mentioned in chapter 5, in class we deliberately include obstacles to understanding in syntax, inappropriate register, tension and emotions, terminological mistakes, lack of textual cohesion and resulting non-sequiturs (this last feature arises from translation mistakes or from lack of cohesion in the utterance–response, e.g. replying to the last part of the utterance first). One can also deliberately break the logical cohesion of an utterance, for example by including the beginning of the following chunk of speech, to challenge the students' mnemonic skills.

As mentioned earlier, remembering figures, dates and names is very challenging indeed for the students unless they are taking notes, and we have tried to integrate these exercises into the dialogues. We give the students the translations of the most difficult terms at the beginning of the course, reducing the number of terms supplied as the course progresses. Sometimes non-technical terms (like 'nuances' in business Dialogue 3) that we know from experience are challenging, are also supplied.

Since these dialogues are intended as an educational tool that can be adapted to any language pair, we have done two things that we hope will be useful to the trainer. First, we have divided the passages into chunks by indicating '//' (in the Italian passages the // marks have also been placed in the English translations that follow). These are simply suggestions about where to break down longish passages, if necessary – if the students cannot yet handle long chunks. Those chunks that are very short are meant for beginners. At times, the chunks have been divided not just according to length but also for topic changes, logical units of meaning, seeking affirmation/reply etc., as would happen in natural conversation. Trainers need to adjust this according to their specific needs, be it memory work, practising consecutive interpreting, or to emulate the natural flow of a dialogue: i.e. should the chunks be as long as possible to avoid interruption and imposition by the interpreter, or short to follow the natural entry/break points? The trainers' approach will also be connected to how 'visible' she wants the interpreter to be: less visible in long chunks in an almost dyadic format, and more 'visible' in short chunks in a more natural sounding – but also more invasive as well as time-consuming – triadic dialogue. Finding the right balance between turns and topic change and chunk length is therefore up to the individual trainer. It must also be remembered that some of the short chunks can be very dense indeed. (Furthermore, the different phases of the conversation, and subsequent turn-taking patterns, will change according to the sector-specific discourse.) Varying the lengths of the chunks of speech is ideal for memory training, and the length of the chunks should in any case be increased to coincide with the students' progression.

Second, we have also included information about when the polite form is being used in the Italian passage ('You' for 'lei') and suggested in the English texts when the You form might be used when translating into other languages.

6.1 The business sector

Dialogue 1   Interpreting for an Italian pasta producer

SITUATION: Ms Whitehouse and Ms Smith from Pasta-pronta, a gourmet Italian pasta shop in York, England, have just arrived in Parma, Italy, for a meeting with Mr Norberto, Marketing Manager of Parilla [1], an Italian pasta producer.

Introductory note: This dialogue is not particularly technical and is quite colloquial in register. It can be used to break the ice and to start interpreting practice in one of the first lessons. The interlocutors in this dialogue have different approaches to business: the Italian manager has a corporate approach that differs from that of his British counterparts who, being shop-owners, are used to 'smaller' deals but are equally interested in making a profit. The dialogue starts with a few introductory pleasantries about the trip, and the register here is reasonably informal. This first dialogue has extended comments, many of which apply generally to the following dialogues as well (greetings, length of chunks, politeness forms, gap-fillers, turn-taking, change of topic, register, translating culture-specific terms, deixis, lists, names, figures, and so on).

IT = Italian speaker (Sig. Norberto, Marketing Manager of Parilla)
E = English speaker (Mrs Whitehouse, from Pasta-pronta in England)
Level of difficulty: easy. The main difficulties here are related to cooking expressions.
Terms supplied: none.
The dialogue contains several business terms (work shift, stock market, etc.), but with terms as easy as these we prefer to test them on the students first and see if they can manage to understand the terms from the context and provide an adequate translation or paraphrase.

IT: Buon giorno Signora Whitehouse. Ben arrivata! Vedo che è con la sua assistente, la Signora Smith. Com'è andato il viaggio? *Good morning Mrs Whitehouse, and welcome to Parma.*[2] *I see You are with Your assistant, Mrs Smith. How was Your trip?*

E: Good morning. We had a very comfortable journey from York, thank You.

IT: Ah, che bello. Due mesi fa ci siamo visti a York, nel Vostro bellissimo negozio, ed è un piacere per noi oggi vedervi qui a Parma. *Oh, good! Two months ago we met in York, in Your wonderful shop, and it's a pleasure for us to have You here in Parma today.*

E: Thank You so much, You're very kind. // When we met last May, we discussed the possibility of going into business together. I'm hoping that we'll be able to continue that discussion today.

IT: Dopo il nostro incontro desidero che visitiate lo stabilimento. Sebbene [3] la nostra ditta operi da 30 anni, lo stabilimento di produzione è stato completamente rinnovato di recente. // Come potrete vedere, gli impianti produttivi sono nuovissimi e ogni reparto è dotato di impianti caratterizzati da tecnologie avanzate dotate di software per il comando e il controllo di linee ed impianti produttivi e sistemi di automazione [4]. // In quanto alla produzione, produciamo 300 kg di pasta al giorno, con una gamma di 32 tipi diversi di pasta [5]. *After our meeting, I would like You to visit our factory. Although our firm was set up 30 years ago, our manufacturing unit has recently been completely renovated. // As You will see, the manufacturing plants are brand new, and every department is equipped with advanced technology that provides manufacturing line- and plant-operating software as well as automated systems. // As to production, we manufacture 300 kg of pasta a day, and a range of 32 types of pasta.*

E: We'd very much enjoy visiting it [6]. We've been offering our customers Your sample foods over the last 30 days, and we're very satisfied with the results of this trial period. The pasta business is growing rapidly in York, indeed, all over Europe! However, customers are becoming more and more demanding in their choice of pasta. [7]

IT: Sapevo infatti che avete clienti molto esigenti [8]. Sono anche molto curiosa, però. Quali sono i tipi di pasta preferiti dai vostri clienti? *I know Your customers are very demanding. Just out of curiosity, which types of pasta do they like the most?*

E: They liked your spaghetti most. As you probably know, last year we were using the products of another pasta manufacturer – Lazzi – who's based in Naples and specializes in macaroni. And although his pasta is good, our customers seem to prefer Yours.

IT: Mi fa molto piacere! Come avrà potuto vedere in questo mese di prova abbiamo diversi tipi di spaghetti di diversa grandezza e gusti (per esempio nero-seppia e verde-spinaci) con diversi tipi di cottura. *That's nice! As You will have noticed during this 30-day trial period, we have different types of spaghetti of various sizes and taste (for instance black-sepia and green-spinach) [9] with different cooking times.*

E: Yes, although I'm sure You're aware of the fact that the British have very different eating habits and are not used to eating pasta 'al dente' the way You do; we tend to cook it much longer.

IT: Si, lo so. Per noi é un grande mistero come riusciate a mangiare la pasta scotta! *Yes, I know. How You can possibly eat overdone pasta is a mystery to us!*

E: There's something else I'd like to discuss with You. We're planning to introduce a new line of fresh pastas because fresh pasta and Italian cuisine have become extremely chic in Britain. // In cities like London and Edinburgh it's all the rage [10]. Tortellini and home-made vegetable lasagna have become a must at any self-respecting dinner-party. // I've even found several courses on how to make home-made pasta in Edinburgh! Our customers are beginning to understand that quality and freshness are essential.

IT: Mi fa molto piacere. Questo nuovo atteggiamento da parte del cliente inglese apre grandi opportunità di una nostra ulteriore collaborazione futura, se riusciremo ad ottenere una quota di mercato sufficiente. // Il nostro Ufficio Marketing, che svolge ricerche specifiche sui bisogni dei mercati e il potenziale di vendita dei nostri prodotti in ogni mercato, ha una Sezione anche nel Regno Unito. // Se lo desidera, posso fissarle un appuntamento con loro domani. *I'm very pleased to hear that. This new attitude on the part of British customers opens up many opportunities for future collaboration, if we manage to acquire a sufficient market share. // Our marketing department – which conducts research on market needs and the sales potential of our products in every market – also has an office in the United Kingdom. // If You like, I can fix an appointment with them tomorrow.*

E: Thank You, that would be extremely helpful.

IT: Ora[11] possiamo parlare d'affari.[12] Partirei raccontandole qualcosa della nostra ditta, in modo che possiate conoscerci meglio e, speriamo, avere piena fiducia in noi. Siete d'accordo? [13] *Ok, now we can get down to business. I'd like to start by telling You about our firm, so that You'll get to know us better and hopefully we'll be able to gain Your trust. Would that suit You?*

E: [Indicates agreement through body language.]

IT: Abbiamo nove fabbriche in Italia, e settanta impiegati negli impianti – uffici e fabbriche – qui a Parma. Il nostro giro d'affari annuo è sui 10.000 euro. // Come può vedere dal Catalogo, vantiamo una gamma di venti prodotti, ovvero sedici diversi tipi di pasta. Nell'ultima lettera lei mi aveva informato che intende piazzare un ordine consistente oggi. *We have nine factories in Italy, and 70 employees in offices and factories here in Parma. Our yearly turnover is around 10,000 euros. // As You can see from our catalogue, we have a range of 20 different product categories, and 16 different types of pasta for each category. In Your latest letter You said that You intended to place a big order today.*

E: Yes, that's correct. Are You prepared to offer us a discount if we buy a large quantity of pasta?

IT: Naturalmente, intendo concederle un ottimo sconto per grandi quantità, ovvero uno sconto del 10% su oltre 2,000 confezioni di pasta. Un confezione contiene spaghetti di varie dimensioni. *Of course, I can offer You a good quantity discount of 10% if You order more than 2,000 pasta packages. A package includes a whole range of spaghettis of different sizes.*

E: We'd like to start with 2,000 packages immediately. If Your spaghetti sells well, and we're quite sure it will, we're prepared to increase the order.

IT: Bene, vogliamo parlare di tempi e modalità di consegna? *Great, shall we talk about delivery times and terms?*

E: Yes please. Could You tell us about your shipment methods and Your delivery terms, please. // When can we expect the first consignment to be shipped? And when can we expect the merchandise to arrive? **[14]**

IT: Potremmo per esempio fissare una data ogni mese, ad esempio il primo del mese per la consegna ordini potrebbe andarle bene? *We could for instance fix a date every month, let's say the first day of every month, for the delivery; would that suit You?*

E: Sure; if Your Despatch Department can finalize the details, that's fine with us. Which mode of transport do You normally use to the UK?

IT: Di solito noi operiamo con trasporto su strada, dato che abbiamo un servizio automezzi per le consegne. // Possiamo concordare una data di partenza per i prossimi ordini? Ad esempio, con i nostri clienti di Hannover in Germania ci siamo accordati sulla consegna al primo del mese. // Prestiamo il massimo della cura all'imballaggio. Come può immaginare, con prodotti come la pasta, soprattutto le tagliatelle, è fondamentale che vengano consegnati in ottime condizioni, e che le confezioni di pasta non vengano rovinate o schiacciate. *We normally ship our goods by road, as we have our own truck service. // Can we agree on the starting date for our next orders? For instance, with our customers based in Hannover, Germany, we deliver our goods on the first day of each month. // The utmost care is given to packaging. As You can imagine, products like pasta, and especially 'tagliatelle', have to be delivered in the very best conditions, and the packaging must not be damaged or crushed.*

E: Yes, of course. What about damaged products? Will we be reimbursed for products that are damaged during transportation?

IT: Sì. Abbiamo una copertura assicurativa per il trasporto. Ora vorremmo parlare di modalità di pagamento. E' la prima volta che lavoriamo con voi, quindi siamo costretti a chiedervi di pagare con lettera di credito. *Yes, we have transport insurance coverage. Shall we talk about payment terms now? This is Your first order, and we're obliged to ask You to pay by letter of credit.*

E: No problem, we're used to working with letters of credit.

IT: Si, abbiamo parlato con il nostro agente nel Regno Unito, che ci ha informato che avete un'ottima reputazione. Già al vostro secondo ordine potremo cambiare modalità di pagamento. *Yes, we talked to our UK agent, and he told us that You are credit-worthy. After Your first order, we'll be able to change the payment terms.*

E: Good, that's fine by us. The last company we worked with was very unreliable, which is why we would like to avail ourselves [15] of Your forwarding company. There seemed to be a carrier strike, constantly, and often there would be a discrepancy between the goods we had ordered and those that were delivered.

IT: Ah, purtroppo non si può escludere la possibilità che vi siano degli scioperi. Come sappiamo, Italia è il paese degli scioperi, noi siamo stati colpiti molto dagli scioperi di ferrovie e linee aeree, per esempio; anche quando la merce viene fermata in dogana possiamo fare ben poco. // Abbiamo però un'ottima reputazione per quanto riguarda la spedizione delle merci. *Unfortunately we cannot rule out the possibility of strikes. As You probably know, Italy is the land of strikes.[16] We have had many railway and airline strikes in the past. When goods are stopped by customs there is not much that we can do about it. // Our company, however, has a very good reputation when it comes to the shipment of goods.*

E: Well, that's good news because in our last consignment a number of goods were missing! And the packaging had broken so that the entire consignment was damaged! [17]

IT: Avete restituito tutta la merce? *Did You return all the goods?*

E: Yes, and luckily it was covered by their insurance company although neither the forwarding company nor the suppliers seemed particularly concerned. // But it's very unpleasant when these things happen. We hope to be able to rely on You.

IT: Vedrà che con noi si troverà molto meglio, ne sono sicuro. *I'm sure You'll be much more satisfied with our company.*

E: I have no doubt we will! Goodbye Mr Norberto, it's been a pleasure doing business with You.

IT: Arrivederci. *Goodbye.*

Commentary

[1] Many of the names are well-known brands that have been slightly modified, partly to contextualize and partly to make it entertaining. 'Parilla' is a pun on the well-known Italian pasta brand 'Barilla'.

[2] Especially at the beginning of the course, students find it difficult to translate 'small talk' greetings and talking about the weather or the journey. These 'openers' are important to break the ice and create a natural progression in the conversation. They also provide register variation and culture-specific translation challenges. It is also a good idea to immediately introduce the use of the You/you forms and the use of given names/ surnames/ titles.

The English translations of many of the gap-fillers, discourse markers and utterances that have a more marked communicative, than semantic, function are not direct translations. Rather, we have used corresponding expressions that we believe would be more natural in English. This is an excellent starting point from which to discuss the pragmatic function of words and expressions, and especially of politeness markers, and to give corresponding examples of these in L1 and L2. Practically any opportunity can be used to show how many different solutions there are to solve grammar- or syntax-related problems. The simplest and clearest solution in the target language is usually recommended, but other options should be discussed too.

[3] The Managing Director is using a quite formal register here. Differences in register between the two languages should be pointed out to the students and the strategies needed to translate them should be examined.

[4] Lists of materials and technical descriptions allow trainers to test the students' memory skills and knowledge of terminology.

[5] In this passage the register rises slightly and the terminology becomes more field-specific and less general. At this point we introduce two numbers, and later more will be introduced. We find that the students tend to forget the first sentence as soon as they hear the numbers, because of an automatic mnemonic process whereby they focus on the most difficult part to remember, the numbers. We vary the length of the chunks according to the level of difficulty that we feel is appropriate to the student's competence at that particular time. This passage can be divided up according to the types of skills the students need to practise.

Pasta is an Italian term that needs no translation. Here, one might list the sectors where source language words are used even in English (music, art, cuisine, etc.) and stress that translating them would not be correct. The same applies to the English terms that are not translated into other languages, for example IT terms such as *mouse*, or words like *governance*.

[6] The reference 'it' is hard to track here because it refers to the beginning of the previous speaker's utterance. Here the trainer can either keep the structure to increase complexity, or divide the previous chunk.

[7] This chunk is quite long, but it is the same line of thought, and stopping here would make it even more difficult for students to remember the whole chunk. It is advisable to use it whole, and to encourage students to interrupt when they need to.

[8] At the beginning, students tend to use more formal, and often complicated, forms in English. They are encouraged to always try and use the simplest form in English rather than a literal translation.

[9] There are potential terminology problems here. Trainers can expand on cooking terminology problems at the end of the lesson. Students can be asked to prepare a glossary on cuisine and to translate a menu. Watching excerpts from British TV cooking programmes is useful also because they hear English regional accents and improve their comprehension skills.

[10] Similar colloquial expressions can be examined. Later, trainers can discuss topical issues like fads or newly emerging habits in British society, like eating good Italian or French wines and typical Italian food. This gives them a chance to engage students in conversation. Students may also work in groups where one of them reports on a local typical product or dish and is interpreted by his colleagues.

[11] Again, gap-fillers, linking words, conversation openers are ubiquitous in, and crucial to, the natural flow of conversation. The forms, functions and levels of formality vary a great deal between languages and are well worth exploring with the students.

[12] Trainers can give students homework assignments such as drawing up a glossary, as more technical terms start to appear: turnover, range of products, placing an order, etc.

[13] Formal turn-taking modes can be discussed, as the Italian interlocutor is suggesting a change of topic. Politeness markers can be highlighted here too.

[14] There is actually quite a lot of material packed into this short chunk for both interlocutor and interpreter to remember. The chunk could be divided into two or even three pieces to make it easier.

[15] A good opportunity to discuss register: 'avail oneself of' versus 'use'. The next sentence is also quite formal, but useful because it is a way to hedge, to avoid a potentially face-threatening criticism.

[16] Here we find another cultural stereotype, which provides yet another opportunity to discuss politeness markers, especially as the speaker is not being deliberately impolite.

[17] At this point it is 'safer' to criticize something that happened in the past with another interlocutor. The tone of voice changes and the hint of tension is dissolved; the atmosphere is decidedly amicable and one gets the feeling that they are closer to reaching an agreement.

Dialogue 2   Starting a business collaboration

SITUATION: An American delegation from World, a software producer, is visiting the headquarters of Info.com, an Italian firm based in Bologna, in Italy, in the hope of establishing a collaboration.

IT = Italian speaker (Sig. Marchi, Marketing Manager of Info.com)
E = English speakers (Mr Smith and Mr Morley, World Export Managers)
Level of difficulty: easy. Some business terminology is used, but no technical terms are mentioned.

This dialogue can serve as an introduction to more difficult ones to follow, where more specific business terms and technical terminology can be used.

Terms supplied: SME/Small and Medium-sized Enterprise (*piccole e medie imprese*).

IT: Buongiorno e benvenuti. Sig. Smith, desideriamo ringraziarla molto di essere venuto a trovarci con il collega, che presumo sia il Sig. Morley. Sono il Dr. Marchi, Direttore Marketing. *Good morning and welcome. Mr Smith, we'd like to thank You for coming to visit us with Your colleague; I assume he's Mr Morley? My name is Mr Marchi, I'm the Marketing Manager.*

E: (Smith)   Thank You. Yes, let me introduce Mr Morley, our Export Manager. [1] I was the one who replied to Your letters and enquiries, but Mr Morley will be responsible for the contract that we'll hopefully sign at the end of our stay in Bologna.

IT: Piacere, Sig. Morley e benvenuto anche a lei. // Effettivamente abbiamo avuto due mesi di intense trattative, prima al telefono e poi per posta elettronica. A proposito, il/la Sig./ra X, la nostra interprete, che avete già sentito telefonicamente e che vi presento, sarà con noi per tutto il vs. soggiorno.[2] // Vi ringraziamo anche perché avete accettato il ns. invito nonostante i recenti scioperi e il fatto che l'aeroporto di Bologna è stato chiuso il 1° maggio e rimarrà chiuso fino alla fine di luglio. *Pleased to meet You Mr Morley, and welcome. // We've been negotiating intensively for two months, haven't we? First on the phone and later by e-mail. By the way, let me introduce Ms/Mr X, our interpreter, to You. You've spoken to her/him on the phone already and (s)he'll be with us for Your whole stay. // We also thank You for accepting our invitation notwithstanding the recent airline and train strikes and with Bologna airport being closed from the 1st of May till the end of July.*

E: Well, thank You very much for inviting us. Yes, we decided to risk it! We felt it was crucial to come now and not cancel the meeting, as we were encouraged by Your interest in our products.

IT: Sì Sig. Smith, da quando abbiamo potuto vedere i vs. prodotti software alla Fiera Informatica di Londra, lo scorso gennaio, volevamo instaurare una collaborazione commerciale con voi. Pensiamo che ci sia un ottimo rapporto qualità – prezzo nella Vs. produzione. *Good, Mr Smith. When we saw Your software products at the Computer and IT trade-fair in London, last January* [3]*, we hoped to start collaborating with You. We think Your products provide an excellent price–quality ratio.*

E: That is indeed very encouraging. Did You examine our new range of products in the catalogue that we sent You a month ago?

IT: Sì. Siamo molto interessati al programma di contabilità per le piccole e medie imprese, ovvero l'ultima versione del prodotto che avevate presentato in fiera. *Yes. We are really interested in Your SME accounting software, the latest version of the software that You launched at the trade-fair.*

E: Well, that is the brand-new version, of course, and only at the end of the year will we be able to draw more definitive conclusions about how viable it might be on the market and when to launch it. I'm happy to say that the results we've obtained so far are very encouraging.

IT: Noi abbiamo svolto un sondaggio presso i nostri clienti ed i risultati, resi noti all'inizio di giugno, ci confermano che il vostro prodotto è in grado di soddisfare le nuove necessità del mercato italiano.[4] *We've conducted a customer survey, and the results – published early in June – confirm that Your products can actually meet the Italian market's needs.*

E: I'm glad to hear that. The SME Accounting Solution proved to be very successful in the UK. I'd like to know if You agree on the price that we quoted in our latest offer.

IT: Il prezzo è un po' alto, e desideriamo discuterne ora. Data la nostra vasta clientela, siamo certi di fungere da motore promozionale per i vostri prodotti, e siamo fiduciosi che vorrete accordarci un prezzo migliore. // Se tutto andrà secondo le nostre previsioni, non escludiamo la possibilità di diventare l'importatore italiano dei vs. prodotti. *The price is a bit high* [5]*, and we'd like to discuss this now. As we have a large customer base, we are confident that we will be a good* *marketing platform for Your products. We are sure You will be willing to offer us a better price. // If everything goes according to plan, we hope to become Your exclusive Italian dealer.*

E: That's an interesting proposal. Let's discuss prices in further detail. Mr Morley, our Export Manager, is at Your complete disposal. The floor is Yours, Mr Morley.

E: (Mr Morley)   Yes, thank You. I share Your view and we are happy to say that we can grant You quantity discounts if You intend to place very large orders. I understand Your survey results are quite good.

IT: Esatto. Ogni nostra decisione commerciale, soprattutto in momento di crisi economica come questo, deve essere supportata da dati di mercato. La nostra ditta è riuscita a superare momenti difficili, e ovviamente dobbiamo rispondere al nostro Consiglio di Amministrazione. // Si tratta quindi di una scelta ponderata e saremmo molto lieti di potere formalizzare un accordo con voi, se possibile, al termine del nostro incontro di lavoro. // Quindi propongo di dare la possibilità al Dr. Morley e al nostro Dr. Bianchi di discutere la bozza di accordo nei particolari. *Exactly. Every business decision we make has to be supported by market survey data, especially during economic crises like the present one. Our company has been able to overcome some pretty tricky financial challenges. Obviously, we will also have to ask for our Board of Directors' approval. // We've examined this proposal with great care and we would be really glad to reach an agreement with You, if possible, at the end of our business meeting. // So I propose we give Mr Morley and Mr Bianchi the opportunity to discuss our draft agreement in further detail.*[6]

E: (Mr Smith)   Sure, we'd be very pleased to let them discuss our agreement. Let's let them work for a couple of hours and then resume our talks in the afternoon. I wouldn't mind walking round Bologna for a couple of hours. I'd love do a bit of sight-seeing in Your beautiful city!

Commentary

[1]   Positions and qualifications can be difficult for students (see Dialogue 1). Getting them to list various corporate tasks and the relevant terminology – Board of Directors, Managing Director, CEO, etc. – is good practice. Whereas in Italian, 'signor', 'signora' and 'signorina' are commonly used, in English one would tend to use the full name, personal name and surname. To make the dialogues more consistent, however, we have generally kept the 'Mr', 'Mrs' and 'Ms' format, but trainers can replace these with first names if needs be. Also, in Italian, 'Dottore', 'Dr.' is used as a term of respect for anyone who has an undergraduate degree (or if it assumed that they have a degree). In the

English translation this could either be dropped and replaced with the first name, or replaced by their professional title, if it is known.

[2] Introducing the interpreter is good practice, in all senses of the word, and gets the students used to being more proactively engaged in the communication, helping them to feel more part of the group and take responsibility. In terms of organizing the conversation, however, it can be tricky, so this needs to be planned ahead of time and the students warned that they will be drawn into the conversation.

[3] Students should be familiar with the major trade-fairs held in their country.

[4] Readers familiar with Italian will note how the register in Italian is often much higher than in English. It is good practice for students to try various translations: maintaining the register or producing a more natural-sounding register for the context.

[5] Negotiating practices vary in different cultures, and students should be aware of the cross-cultural issues they may be facing. Delicate phases such as this also require an uninterrupted flow of conversation and one should be careful not to break up long chunks, creating undesired interruptions that might have a negative effect on the other interlocutor's image of the speaker.

[6] This is a very important passage in the conversation, the final phase of the preliminary negotiations, and should be handled with care at the interpersonal and stylistic level because the objective is to try to obtain the best deal for both parties without ruffling too many feathers and by presenting oneself as reliable, serious and trustworthy (hence the emphasis on empirical data). Thus, politeness strategies, tone of voice, body language, completeness of message/topic are all important features.

Dialogue 3   Interpreting for the leather goods industry

SITUATION: A well-known fashion company, Pandarina Duck, has invited the English designer David McNamara to visit them in Bologna. Mr Rossi, Department Manager of 'New Trends', welcomes him. They have been collaborating for a long time and know each other well. Note that the Italian speaker uses the non-polite form of 'you'.

IT = Italian speaker (Mario Verdi, from 'Pandarina Duck')
E = English speakers (David McNamara, designer)
Level of difficulty: medium.
Terms supplied: leather goods (*pelletteria*), shade/tone, nuance (*sfumatura*).

IT: Buongiorno David **[1]**, e ben arrivato. Come è andato il viaggio? *Good morning David, welcome. How was your trip?*

E: Thank you Mario, the flight was fine. It just took us a long time to get here from the airport; there was a terrible traffic jam!

IT: E' in corso una fiera, quella del settore edile, ed è una fiera molto importante. // Pensa questa fiera è la seconda in Europa. Bologna è ormai diventata un importante centro fieristico europea. // Si tengono due fiere edilizie l'anno, ed attirano molti espositori, non solo dall'Europa, ma da tutto il mondo. La nostra sede purtroppo è ubicata molto vicino alla Fiera **[2]**. // Vieni, sediamoci nella sala riunioni ed aspettiamo gli altri. // Vuoi bere qualcosa? Magari un buon espresso? No, temo che tu preferisca un tè. *There's a trade-fair on now, dealing with the building sector, it's a major event* **[3]**. *// This trade-fair ranks second in Europe, just imagine, and Bologna has now become a major European exhibition centre. // Two building sector exhibitions are held every year, and they attract many visitors, not only from Europe, but from the whole world. Unfortunately our headquarters is located very close to the fair. // Come this way, let's go and sit in the meeting room and wait for the others there. // Would you like something to drink?* **[4]** *Maybe a good Italian espresso coffee? No, I suppose you'd rather have a cup of tea?* **[5]**...

E: Well...., actually, could I have a latte? **[6]** It was a tiring morning, I had to get up very early to be at Heathrow at 8.

IT: Un caffelatte? Va bene.... Mando Cinzia al bar qui sotto. // Sono arrivati tutti, e alcuni li vedrai per la prima volta. Ora te li presento. Ecco Ornella Bianchi, la nostra esperta colori, Mimma Furlo, reparto valigeria, e Toni Cantaro, della produzione. // Fra un'ora arriverà Massimo, che già conosci bene. *Milk and coffee? Well, ok… I'll send Cinzia to get one from the bar downstairs. // They've all arrived –some of them you haven't met before. Now, I'm going to introduce them to you. Here's Ornella Bianchi [7], our colour expert, Mimma Furlo, from the Leather Goods Department [8] and Toni Cantaro, from the Production Department. // Massimo, you know him well – will join us in an hour.*

E: Oh yes, we've been collaborating closely over the last few months. Will I meet your Managing Director?

IT: Certamente, come d'accordo lo vedrai subito dopo questa riunione, verso le 13, così poi andrete insieme a pranzo. Abbiamo prenotato in un ottimo ristorante in centro. *Of course. As we agreed, you'll be meeting him immediately after our meeting, at about one o'clock, and then you'll go out for lunch together. We have booked a table for you in an excellent restaurant in town.*

E: That sounds great. Ah, these are the new bags, I really like them.

IT: Ci fa molto piacere. Come ti dirà Mimma, abbiamo avuto molte difficoltà per trovare la pelle più adatta al modello da te disegnato. Credo che anche Ornella voglia parlartene. Mimma, a te la parola. *Oh good, I'm glad to hear that. As Mimma will tell you, we had quite a lot of trouble finding the right leather for the model you designed. I think Ornella would also like to discuss this with you. Mimma, you have the floor [9].*

IT: Sì, David, quella sfumatura di verde ci ha causato non pochi problemi! Ti ricorderai che l'ultima volta che sei venuto a Bologna abbiamo esaminato insieme la gamma di colori ed abbiamo stabilito alcune sfumature. Poi però adattare tipo di pelle e sfumatura di colore ci ha dato del filo da torcere. *Yes, David, that shade of green caused no end of trouble! You probably remember that the last time you were in Bologna we examined the new range of colours together and we decided on a few nuances. [10] Later on, though, we had such trouble combining the right type of leather and colour shade.*

E: Sorry, but I see you've done a great job. [11] That's exactly what I had in mind when I sent you my latest design. Great! // The small size is even better, I think. May I have a look at the new wallets? [12]

IT: Eccoli qui. Questi sono in vitello, e siamo molto contenti del risultato. Speriamo tu sia d'accordo. *Here they are. They are calfskin [13], and we're very satisfied with the outcome. We hope you'll think so too.*

E: Oh yes, I really appreciate the work you've put into this. Well done!

IT: Prima di passare all'abbigliamento, e di chiedere a Monica di indossare gli abiti dell'ultima collezione creata in base ai tuoi disegni, vorremmo mostrarti le nuove valigie, anche se non sono di tua competenza. Cosa ne dici? *Before passing on to garments, and before asking Monica to try on the outfits in our latest collection based on your designs, we'd like to show you our new suitcases, even though it's not exactly your area of expertise. What do you think?*

E: I'd love to have a look. I met the French designer who created them and I think he's very good. Look, they're both stylish and practical too.

IT: Noi siamo particolarmente contenti del nuovo trolley. Come vedi è di un'estrema praticità. Ha una linea molto semplice, che permette di chiuderlo in questo modo, per ridurre al minimo l'ingombro. *We're particularly satisfied with our new trolley-suitcase. As you can see it is extremely practical. It has a very simple line, and you can close it like this so that it takes up very little space.*

E: Mmmm. It's really an excellent new collection. When are you planning to launch it on the market?

IT: Come sai, ora stiamo lavorando sodo per essere pronti per Pitti Moda Uomo. Poi, ci saranno le giornate della 'Moda valigeria' a Milano, e vorremmo essere pronti per lanciare la nuova collezione in quell'occasione, se ci riusciamo. // Ora David, siamo pronti per mostrarti i capi dell'ultima collezione da te ideata; se sei d'accordo, naturalmente. *As you know, we've been working very hard to be ready for the Pitti fashion show for men, Pitti Modo Uomo [14]. After that, there's the 'Moda valigeria', the suitcase exhibition in Milan, and we'd like to be ready to launch our new collection on that occasion, if possible. // Now, David, we're ready to show you our new collection, the last one that you designed, if that's ok with you...*

E: Yes sure, that's fine with me.

IT: Benissimo; se ti siedi puoi vedere Monica quando esce dalla sala prove ed entra nel salone grande. *Great. Have a seat, why don't you; from here you'll be able to see Monica coming out of the fitting room into the large hall.*

Commentary

[1] Again, greetings and leave-takings can be discussed at this stage. Discussing a variety of registers is useful.

[2] This chunk is long, but by the end of the course students should be able to interpret a passage of this length without asking for repetition.

[3] Or 'there's an important building sector trade-fair on right now'. Practically any excuse is good to discuss differences in syntax between the language pair(s) you are using and whether or not it is better to follow the original

syntax when grammatically possible, or reduce syntactic redundancies to provide a more succinct rendition in the target language – not just for clarity and to ease comprehension for the listener, but also to reduce the time of delivery.

[4] 'Let's go', 'come', let's sit there', 'sit down', 'have a drink', 'have a cup of tea/coffee' ('drink a cup of tea'), 'would love/like a cup of tea' etc. are extremely useful everyday expressions that, although absolutely trivial for native speakers, can be difficult to get right for so many students. Expressions like these are important to signal interpersonal relations between the interlocutors by indicating degrees of formality. If the students manage to get them right they also do a great deal to improve the quality of their L2 production.

[5] Students should learn to be confident, relaxed and natural-sounding when translating pleasantries. They should be reminded that intonation and body language are very important here. Humour is also a part of the exchange of pleasantries, although, especially when there is a hint of irony or 'pretend' disapproval, it is a minefield for cultural misunderstandings.

[6] Indeed, the *cup of tea* is based on the stereotype of the English as a tea-drinking nation, which probably isn't as true of present-day Britain as it might have been in the past, not least since the advent of Starbucks and (supposedly) 'Italian coffee shops'. David's reply achieves a double effect. Mario thinks he's accommodating David by offering him 'English' tea while David thinks he's being 'Italian' by looking for the non-existent 'latte', the result being confusion and misunderstanding.

[7] This is a good passage to practise names in L2, but also in L1, which are difficult to remember and require good note-taking skills. Students can use two strategies here: either asking the speaker to interrupt after uttering each single name and the relevant position, or taking many notes.

[8] The different departments of the company include some challenging technical terms. Whether or not to supply these terms in advance is up to the trainer. The advantage of not supplying them is that under stress and time pressure the students are encouraged to find a viable solution spontaneously or to ask for explanations.

[9] This might be bit formal, but 'floor terminology' could be discussed here ('to give the floor to someone', 'let's do a round of the table', 'what do you think, Mimma?', 'is there anything you would like to add, Mimma?', etc.). Also, terminology such as *agenda*, *Chairman* and *Chairperson*, *panel, panellist, question and answer time*, etc. could be discussed here.

[10] Foreign loanwords for L2 speakers may cause comprehension problems, as they are pronounced using English phonetics.

[11] Even though the difficulties encountered are reported, the designer did not get offended. Trainees should be aware of, and wary of, taking on a role around conflict preventer.

The combination of colloquial conversational register and technical terms is good practice.

[12] A possible point at which to discuss cross-cultural aspects of body language, proximity, etc.

[13] Translating technical terms is always a problem. Getting hold of a company's catalogue showing the different types of leather products is useful. Here, students might compile a fashion terminology glossary, and visit the main fashion houses' original websites to find terms and words in context.

[14] The name of this fashion show is an opportunity to discuss how to translate names of well-known events. The original name can often be kept, followed by the target language version, if the students know it; alternatively, the meaning of the original can be paraphrased or explained.

Dialogue 4 Interpreting for the Swatch company

SITUATION: A Swiss delegation of the well-known watchmaker Swatch is visiting Verona, Italy. The purpose of this visit is the acquisition of 'Tempo Prezioso', an Italian manufacturer of high-quality watches who is going out of business.

IT = Italian speaker (Sig. Rossi, from 'Tempo Prezioso')
E = English speakers (Mrs Schaffner and Ms Singh, from the 'Swatch' company)
Level of difficulty: medium-to-difficult. There are several business terms, not necessarily highly technical but often requiring an explanation, a lot of foreign names, and the chunks to be translated are quite long. Figures and names of people and towns add to the level of difficulty.
Terms supplied: none.

IT: Buongiorno e benvenuti! Signora Schaffner, che piacere vederLa ancora! Ringraziamo innanzitutto Lei e i suoi colleghi di avere accettato il nostro invito e di essere venuti in Italia invece di incontrarci a Syracuse. // Speriamo di potervi offrire un'esperienza culturale e culinaria piacevole, e di utilità per entrambi. // Le presento la Signora x, la nostra interprete, che rimarrà accanto a voi per l'intero soggiorno. *Good morning and welcome! Mrs Schaffner, I'm very glad to see You again. First of all let me thank You and Your colleagues very much for accepting our invitation and for having come to Italy instead of meeting us in Syracuse. // We hope Your stay will be profitable as well as pleasant, from a cultural and culinary point of view too. // I'd like to introduce Ms/Mr .... (name of the student) to You. (S)he is our interpreter, and will be with You throughout Your visit.*[1]

E: (Mrs Schaffner) Thank You so much. Yes, we're very much looking forward to our stay here. We're hoping to see a concert at Your beautiful open-air theatre [2]. Most of the team here are both history and art lovers, so we see this as an opportunity to indulge in our hobbies! And naturally we're looking forward to tasting Your celebrated food and wine...

IT: Il volo ha avuto ritardi? *Was your flight late?*

E: Not really, only half an hour, which is not bad, mainly due to the long security checks at the airport, but we didn't have any other problems.

IT: Benissimo. Nel nostro ultimo incontro abbiamo discusso la situazione qui in Italia; purtroppo il mercato di articoli di lusso è in crisi in tutta Europa. *Inoltre la situazione politica mondiale degli ultimi due anni non è stata favorevole. Excellent. When we met last we discussed the situation here in Italy. Unfortunately the market for luxury items is declining all over Europe. Moreover, the world-wide economic situation over the last two years has not helped either.*[3]

E:　Yes. Our top management was anxious for us to meet as soon as possible to come to an agreement about the acquisition of Your company. I'm glad the take-over bid went so smoothly and swiftly; there are still a few administrative things to be taken care of, however. These things can be so time-consuming! **[4]**

IT:　Eh, è proprio vero. So che dopo aver acquisito la nostra ditta sono molte le decisioni da prendere. Siamo stati sempre in contatto negli ultimi mesi, ma siamo a vs. completa disposizione se avete bisogno di chiarimenti. *How true... I know that after purchasing our company there are a lot of decisions that will need to be taken. We have constantly been in touch over the last few months, but we are at Your complete disposal if you need any clarification.*

E:　We still have many things to discuss! First of all, however, let me introduce my team to You. // This is our Marketing Manager, Ms Singh, our Italian Section Coordinator, Mr James, our Legal Representative, Mrs Bell, and our Export Manager, Mr Hiroshi. **[5]**

IT:　Buongiorno Signora Singh, Signor James, Signora Bell, Signor Hiroshi. E' un grande piacere per noi potervi ospitare qui. Vi prego di dirmi quali sono gli aspetti che desiderate discutere oggi. // Prima, però, vorremo chiedere se potete parlarci della vostra ditta. Ovviamente, una ditta famosa come la vostra non ha bisogno di introduzione, ma ... *Good morning Ms Singh, Mr James, Ms Bell, Mr Hiroshi. It's a pleasure for us to have You all here. Could you please tell me which issues you'd like to discuss today? // Before doing that, however, could you please tell us about Your firm? A well-known company like Yours needs no introduction, of course, but ...*

E:　With pleasure, in fact Ms Singh has prepared a brief PowerPoint presentation to give You a few bare facts about our company. Would you do the honours, Shirin? **[6]**

E:　(Ms Singh)　Of course! As You know, Swatch is one of the leading watch manufacturers on the world market, and our most famous product is the Swatch watch. The Swatch group consists of twelve companies, including Tissot, Omega and Swatch, with a head office in Zurich, where we employ ninety people. We have nine factories in Europe and another main office. **[7]** We're constantly looking for bright young designers to recruit – our head-hunters [8] in the research and development division have a full-time job!

IT: E per quanto riguarda la produzione? Vuole aggiungere qualcosa? *And what about production, is there anything You would like to add?*

E: (Ms Singh)   In order to keep up production, our factories never close, we're open 24 hours a day and we work three 8-hour shifts.[9] // We have a daily production of 25,000 watches with significant profits every year – annual sales total more than $20 billion. As You all know, of course, we produce affordable watches – in fact, our prices have remained unchanged since 2008!

IT: So che presentate una nuova collezione di orologi ogni anno..... *You present a new collection of watches every year, don't You ...?* [10]

E: As You can see on the next slide, every year we design a new collection which is usually presented in spring. There are eighty products in our catalogue. We launch the new collection on the market at the same time all over Europe, the US, Japan, South Asia and South America. // We follow the fashion scene very closely, of course, and work in close contact with the major fashion houses.

IT: Grazie infinite, Ms Singh, è stata una presentazione molto chiara. *Thank You so much, Ms Singh, Your presentation was very clear.* [11]

E: (Mrs Schaffner) Thank You, Ms Singh. One of the points we'd like to discuss with You is the future marketing [12] and distribution of Your product range. Naturally, the marketing strategies we employ for our youthful, lively and affordable Swatch watches will have to be adapted to a much more demanding and resourceful target market. [13] // The sales organization also is different from Yours; eventually You'll have to tell us a bit about Your sales distribution.

IT: Con piacere. Noi abbiamo sempre venduto principalmente alla piccola distribuzione, ovvero piccoli negozi di orologeria e gioielleria, ad alcuni grandi magazzini di alta qualità, come Harrods a Londra, Saks Fifth Avenue a New York e anche di medio-livello come Printemps a Parigi [14]. *We'd be delighted. We have always sold our products mainly to small watchmakers' and jewellers' shops, some luxury department stores, like Harrods in London, Saks Fifth Avenue in New York and even medium-segment ones like Printemps in Paris.*

E: We have a global market as you know, but what about Your market niche?[15]

IT: Come sapete abbiamo un'ottima rete di vendita a livello europeo, soprattutto Italia, Francia, Germania e Gran Bretagna. Da cinque anni abbiamo anche aperto negozi negli Emirati Arabi, in Arabia Saudita, e in Kuwait [16]. // Riteniamo che il mondo arabo sia un mercato emergente molto promettente; certo che la crisi mondiale politica potrebbe cambiare la situazione.... Vediamo un nuovo mercato anche in Giappone, Singapore e Malesia. *As You know, we have an excellent sales network throughout Europe, mainly in Italy, France, Germany and Great Britain. Five years ago we also opened some shops in the United Arab Emirates, Saudi Arabia and Kuwait. // We believe the Arab world is a very promising emerging market. Of course the current world crisis might change things ... We think that Japan, Singapore and Malaysia are also interesting new markets.*

E: Let's talk a bit about marketing. We loved your last marketing campaign! [17]

IT: Ci fa molto piacere che vi sia piaciuta. In effetti [18] ci siamo rivolti ad una nuova agenzia pubblicitaria, un'azienda giovanile e innovativa. Volevamo cambiare leggermente immagine, e renderla più dinamica. // Dobbiamo dire che la campagna ha avuto un gran successo soprattutto fra i giovani. *I'm very pleased that You liked it. We hired a new advertising company, a youthful and innovative firm. We wanted to change our image a little to make it more dynamic. // I must say that this campaign has been a great success, especially among young people.*

E: I'm so glad to hear that. Let's talk a bit about price: what kind of price range do You operate with?

IT: Partiamo da un prezzo di 2.000 euro per l'orologio meno costoso, e poi andiamo su, fino a 10.000 euro per gli orologi ideati dai designer più noti, e qui mi riferisco alle case di moda dei grandi stilisti. // Ovviamente si tratta di orologi personalizzati, fatti su richiesta del cliente, e spesso sono impreziositi da gioielli. *We start from 2,000 euros for the cheapest watches, and then the price goes up to 10,000 euros for the watches created by famous designers – and I'm talking about the big fashion houses. // Of course, we are talking about custom-made watches, often with inset jewels.*

E: Who are Your main competitors?

IT: I nostri concorrenti principali sono Cartier e Rolex. Tutti sentono l'impatto della crisi economica mondiale, anche se forse meno del previsto. // La Borsa é stata colpita duramente e ovviamente ci vorrà molto tempo prima che la situazione si riequilibri. Riteniamo che alcuni dei nostri concorrenti saranno acquisiti dalle società più grandi, come siamo stati noi. *Our main competitors are Cartier and Rolex. They are all bearing the brunt of the international economic crisis, although maybe less than was expected. // The stock market was badly hit, and naturally it'll take a long time to recover. [19] We believe that some of our competitors will be taken over by bigger companies, as we were.*

E: What are Your payment terms? Do You always require cash on delivery?

IT: Come potete immaginare, i nostri clienti non hanno di solito problemi di liquidità, ma ogni tanto può capitare che un giovane imprenditore possa necessitare di una proroga dei termini di pagamento. In questi casi, dopo averne controllato la situazione finanziaria, concediamo loro le proroghe richieste. // Vorrei anche sottolineare che spesso i nostri clienti sono rappresentanti della mondanità, del *jet set* o dell'aristocrazia, e i nostri agenti capiscono benissimo quanto sia importante la discrezione in questi casi. *As You can imagine, our clients do not normally have cashflow problems, but sometimes a young entrepreneur may need to extend the terms of payment. When this happens we check their financial situation and then give them the extra time they need to pay. // Let me also add that our customers often come from the jet set or the aristocracy, and our agents understand how important confidentiality is in these cases.*

E: And what about delivery terms?

IT: Naturalmente, per i pezzi più preziosi richiesti sul'ordine del cliente, ci vorranno alcuni mesi. Per i pezzi più standard, di solito riusciamo a consegnarli entro tre settimane se non gli abbiamo in magazzino, ovviamente. Tutti i pezzi sono assicurati fino alla consegna al cliente. *Naturally, with the most valuable items made to order, it will take a few months. As to the more standard items, we normally deliver them within three weeks, if they are not in stock, of course. All of our items are insured up to delivery to the customer.*

E: Naturally...

IT: Adesso il nostro Amministratore Delegato Vi aspetta, così avrete la possibilità di discutere con lui di tutte le questioni principali e delle vostre prospettive future. *Now our Managing Director is waiting for You. You can discuss any important issues You think need addressing and prospects for the future.*

E: We're looking forward to seeing him!

IT: Benissimo, oltre tutto parla benissimo l'inglese, vista la sua lunga esperienza lavorativa in Inghilterra. // Poi vi porteremo in un ottimo ristorante veneto vicino all'Arena. Per adesso, grazie mille della visita e arrivederci. *Excellent. He speaks English well because he has worked in the UK for a long time. // Afterwards, we will have lunch at a very good local restaurant close to the Arena. Thank You so much for visiting us. I hope to see You soon.*[20]

Commentary

[1] This is a good place to discuss introductions and the need for repetition of unfamiliar and/or foreign names to avoid mispronouncing them (see note 5). Again (see Dialogue 2), this passage also shows us how the student-interpreter can be introduced into the conversation; this should be done carefully however because it can create a great deal of confusion when too many roles are being played out at the same time. It is a good opportunity to discuss the use of the 1st and 3rd person (should the interpreter refer to herself in the 3rd person?) and logistical and practical aspects such as where to sit/stand.

[2] Reference is made to the Arena in Verona here. The art and cultural heritage of a country is often referred to in conversations such as these, and translating names of places and people can be tricky. Students should be familiar with the major monuments in their own and in their L2 countries.

[3] Here we move into the next, more technical, phase of the meeting. Being informed of the current political and economic situation often helps the students grasp the meaning of conversations regarding current affairs.

[4] If possible, this chunk should not be divided. When chunks are quite dense and beginner-level students might have problems, then the trainer should speak slowly and articulate clearly.

[5] Interpreters should always ask for the list of participants' names and positions in advance to avoid mispronouncing, or offending anyone. Foreign names always pose problems. The students should learn to mimic sounds and practise understanding and repeating foreign names.

[6] An interesting expression to discuss and to translate. When many different interlocutors are involved in the conversation it is important to use body language and gesture to keep track of who is who! This is also an opportunity to discuss the use of first names in everyday conversations (especially prevalent in US English) to create familiarity and/or an assumed bond.

[7] It might be a good idea to try and read the whole chunk to students, and stop only if they ask you to. It is also a good way to force them to interrupt when they realize that they won't be able to remember the whole chunk.

[8] 'Head-hunters' has always been a challenge for our students, but a good way to introduce new terminology.

[9] In this passage the main difficulties are due to figures, which should be taken note of. The word '*shift*' was unexpectedly a challenge but one which could be understood from the context. Again, this dialogue is packed with information and in passages like this, with both technical and semantic (colloquial) challenges, the text should be read slowly and articulated carefully.

[10] At some point the trainer could talk about the ubiquitous and so useful 'tag questions' in English and their pragmatic and interpersonal functions, which are expressed differently in other languages.

[11]   Compliments can be very difficult and are also highly culture-specific: students should learn how to praise interlocutors using the most appropriate form and register in the target language, without exaggerating or imposing, which might have the opposite effect. The culture-bound aspects of 'positive and negative face' could be raised here.

[12]   In the Italian language there is a special marketing jargon which retains many of the original English words, which are often mispronounced by Italian speakers. Students should be made aware of this. We suggest they use the corresponding Italian term whenever possible, to widen their Italian vocabulary along with their English one.

[13]   This part contains several business terms and expressions that can be analysed at the end of the dialogue.

[14]   This is a list of famous department stores that the students should be familiar with (general cultural knowledge).

[15]   When French words are used in English they are often not understood by Italian students, as they are pronounced differently.

[16]   Again, names of geographical locations often pose problems. A possible exercise here is that of specifying the official languages of these countries, i.e. Saudi Arabia – Arabic.

[17]   This comment allows trainers to analyse other marketing terms.

[18]   There are so many 'false friends' between Italian and English, and this is one of them. Rather than saying 'in effect', 'effectively' or 'actually', here it could just be dropped, perhaps finding some other emphasizing marker to compensate. ('As a matter of fact', 'in fact', 'on the contrary', 'instead' are also typical trouble spots, as is the ubiquitous Italian 'welcome' which does not always sound equally natural in English.) Making a list of typical false friends also on the basis of the mistakes that emerge during the role-play that week, is useful because students are under a great deal of stress when interpreting and easily resort to the most obvious-seeming corresponding term. Sometimes the opposite happens, that they avoid similar-sounding correspondences at all costs, even when correct. In due course however they should feel confident enough and internalize the terminology well enough to produce more natural-sounding renditions.

[19]   Again, students should be informed about international economic events. They should be familiar with key economic concepts and terminology. If they know the topic well their task becomes much easier.

[20]   Leave-takings and how to adjourn meetings could be discussed here.

Dialogue 5   Interview with an Italian fresh pasta producer

SITUATION: The British launch of one of Italy's favourite fresh pasta brands, Filippo Brana in London. The founder of the company, the elderly Filippo Brana, accompanied by other managers from the company, is being interviewed by a food journalist.

IT = Italian speaker (Filippo Brana)
E = English speaker (English journalist)
Level of difficulty: difficult. Very specific food terminology is used and reference to food habits in Italy and the United Kingdom.
Terms supplied: Fresh pasta (*pasta fresca*), filling (*ripieno*), truffle (*tartufo*), home-made pasta (*pasta fatta in casa*).
Introductory note: This dialogue takes the form of an interview, which of course has specific discourse features that are different from the earlier dialogues, which are more spontaneous. The orderly turn-taking of the interviewer and the focussed answers of the interviewee facilitate recall, not only because of the short length of the chunks but also because they generally represent a single unit of meaning and do not change topic midway (with some exceptions). The interlocutors pass the turn to each other in an orderly and predictable fashion, again making it easier for the students to provide an accurate rendition (as compared to reproducing fragmented speech and different units of meaning within the same chunk). Chunks can be kept long or short, according to the specific pedagogical requirements. Note that the first chunk from the interviewee and the last sentence of the interviewer function as opening and closing remarks, in the manner of prepared speech. Also, this type of interview format allows for the inclusion of a highly colloquial register that pertains to the private domain (especially in the replies of Mr Brana when he talks about his family's eating habits), as well as humour.

E: We'd like to thank Filippo Brana and his staff for inviting us to the British launch of their products here in London.

IT: E'un grande piacere per me essere venuto a Londra, insieme ai miei dirigenti, per lanciare per la prima volta i nostri prodotti con il nostro marchio nel mercato inglese, davanti a un gruppo di giornalisti specializzati nel settore dei prodotti alimentari e del vino. // Ed è un grande piacere per noi essere in questa una magnifica città. E' un'occasione speciale, crediamo, per riuscire a convincervi dell'alta qualità dei nostri prodotti di pasta fresca.[1] *It is a privilege, for me,*

to be here in London with my managers, to officially launch our products
using our own brand names for the first time on the UK market, before a
group of journalists who are specialized in food and wines. // We are also very
pleased to be in this wonderful city. This is a great opportunity, we believe,
to convince you that our fresh pasta products are of the highest quality.

E: We understand that your tortellini are deliciously thin and amply
filled. Is it not true that your products were sold at Harrods first,
and then at Sainsbury's, using their brand-names?

IT: Sì, ha perfettamente ragione. Prima infatti, i nostri prodotti venivano
venduti con il nome dei grandi magazzini che li vendevano,
Sainsbury e Harrods, ad esempio. Abbiamo pensato che il modo
migliore per convincervi fosse quello di farveli assaggiare. // Quindi
vi presento i miei colleghi: la Dott.ssa Miriam Grani è la Direttrice
delle Pubbliche Relazioni, la Sig.ra Maira Calderoni è la Direttrice
Prodotto, il Sig. Polli è il Direttore Marketing Italia [2], gli altri colleghi sono già al lavoro in cucina. *Yes, exactly. Until now, our products*
were in fact sold with the brand-name of the big department stores that
used to sell them: Sainsbury and Harrods, for instance. We thought the
best possible way to convince You was to let You taste them. // Let me
introduce my colleagues to You: Miriam Grani is our Public Relations
Manager, Maira Calderoni is our Product Manager, Mr Polli is Marketing
Manager for Italy, their colleagues are already working in the kitchen.

E: Yes, we saw ten men and two women behind these pots of boiling
water, preparing today's lunch. Some are cutting Alba truffles [3]
over plates of shiny ravioli, others are pouring balsamic vinegar on
three-year-old parmesan cheese [4].

IT: Sì, è vero, se volete questa è l'*Italian way,* cioè è bello vedere gli alti
dirigenti al lavoro in cucina, per dimostrarvi che conoscono bene il
loro lavoro, e lo amano anche! *Yes, absolutely. This is the 'Italian way';*
it's nice seeing top managers working in the kitchen to show you that they
love what they're doing and also that they know what they're doing!

E: Mr Brana, you are the founder of this company, we've been told
that you launched the company nearly 50 years ago, delivering
pasta on your motorbike. Now you own a chain of restaurants and
it is your happy face we see that smiles from every packet. You also
star in all your commercials.

IT: Sì, è vero, e il video che vi mostreremo ora vi spiegherà la nostra
storia. Preferisco mostrarvi il video piuttosto che parlarvi a lungo. //
Per quanto riguarda la nostra storia vi dico solo che fra due anni la
nostra azienda festeggerà i 50 anni di attività. Sono 50 anni che io
lavoro nel settore della pasta fresca e ne sono felice. *Yes, that's correct,*

and the video you are going to watch now will tell you the story of our company. I'd rather show you the video and not bore you by talking too much. // As to the history of our company, what I can tell you is that in two years time we will celebrate our 50th anniversary. I've been working in the fresh pasta sector for nearly 50 years, which I am very pleased about.

E: Did you work in any other sector before becoming a leader in the fresh pasta business?

IT: Guardi, io sono praticamente nato in una bottega della pasta fresca, o meglio, prima ho imparato a fare il pane e poi a impastare i tortellini. Mio padre era un fornaio, e io ho cominciato a lavorare quando ero ancora molto giovane, nel forno di famiglia. *Well, I was practically born in a fresh pasta workshop.* [5] *That is, I learned how to make bread first, and then tortellini dough. My father was a baker, and I started working when I was still very young, in our family bakery.*

E: Today, you're officially entering the British market; what's your main target, Mr Brana?

IT: Noi prendiamo con grande serietà il mercato inglese. Sappiamo che gli inglesi amano la pasta fresca, e vogliono prodotti di alta qualità. Quasi un quarto dei consumatori inglesi acquista la pasta fresca. *We take the UK market very seriously.* [6] *We know that the English like fresh pasta, and want high-quality products. Almost a quarter of the English consumers buy fresh pasta.*

E: Yes, sales of fresh pasta are now worth £70 million [7] a year.

IT: Sì, e voi mangiate più pasta fresca degli italiani. Gli italiani mangiano la pasta ogni giorno, ma di solito mangiano pasta fresca solo una volta a settimana. *Yes, you eat more fresh pasta than Italians do. Italians eat pasta every day, but they usually eat fresh pasta only once a week.*

E: You are also promoting ready-made pasta, we understand? [8] // How often do you make your own pasta, Mr Brana?

IT: Io non la preparo mai adesso. Lo facevo in passato. Mia sorella, che ha 90 anni, fa ancora la pasta. Il giorno di Natale siamo tutti da lei a pranzo, e lei ne è molto orgogliosa. *I never prepare it anymore. I used to in the past. My sister, who's 90, still prepares her own pasta. On Christmas day we all go to her place for lunch, and she is very proud of that.*

E: And what about the sales and marketing directors, are they good chefs?

IT: Loro cucinano tutti bene, come vedrete fra poco. *They are all good chefs, as you will see in a while.*

E: What about the current CEO, your son Giorgio, how often does he eat pasta?

IT: Lui mangia pasta fatta in casa regolarmente. *He regularly eats home-made pasta.*

E: We examined your surveys attentively and think they are really revealing. Only 5–10 per cent of households in the Emilia Romagna region, Italy's fresh pasta-eating heartland, [9] make their own pasta, and you can be sure the figure is considerably lower in other parts, where dried durum wheat pasta, which no one ever makes at home, is the mainstay.

IT: Infatti, uno dei miei più stretti collaboratori dice sempre, e lo cito: 'Non lo dovrei dire in Italia, dove si è sempre detto che la pasta comprata non potrà mai essere buona quanto quella della nonna, ma è così, la nostra pasta è proprio buona, e questa è la verità. // Noi abbiamo gli ingredienti migliori, e il know-how per produrla in serie. Inoltre, una volta aperto il pacchetto, è pronto in un minuto.' *One of my close collaborators always says, and I quote: 'I shouldn't say this in Italy, where they always say that ready-made pasta is never as good as the pasta your grandmother used to make, but the simple truth is that our pasta is really good. // We have access to the best ingredients and we have the know-how to mass-produce it. And, you just open the packet, and it's ready in a minute.'*

E: Is that such a dirty secret? To the Italians, maybe, but for us I think it is rather reassuring. We are so used to beating ourselves up for being the laziest cooks in Europe, for lacking a culinary heritage and succumbing to convenience, that it's nice to know that our friends on the continent aren't above taking the odd short-cut. // The likes of Jamie Oliver [10] give the impression that an Italian mama would rather put pineapple and sweetcorn in her hair and call herself a Hawaiian than serve up shop-bought pasta. [11] // But now that we know the truth, we can blow the dust off our pasta-rolling machines – and throw them straight in the bin.

Commentary

[1] Note the use of informal 'you'. This passage differs from the utterances in the earlier dialogues in that it seems to be a speech fragment with the discourse features of prepared rather than spontaneous speech, as mentioned in the *Introductory note*. We have divided this chunk into two passages, but it is a good opportunity to practise a slightly longer chunk with a more definite structure, as if it were the beginning of a speech.

[2] Again, this is a good place to practice note-taking, memory and pronunciation with lists of names.

[3] Translating menus and regional specialties can be very difficult. They are, however, useful because business negotiations often continue over a good

meal. Moreover, these terms could be given to the students as homework: describing special dishes of the region they come from, and discussing an L2 translation of all the relevant names and cuisine-related terms.

[4] The names of some ingredients should be maintained in Italian or a literal translation of the same is used, i.e. *Parmigiano* = Parmesan cheese, *aceto balsamico* = balsamic vinegar. Discuss other difficulties in translating menus, like having different names for an animal and its meat, i.e. *deer/venison*. Trainers can expand on cooking terminology problems at the end of the lesson.

[5] There is a seeming lack of cohesion (self-repair) between these two sentences (two clauses in the Italian), which is a good opportunity to discuss 'natural errors', inference and implicature in spontaneous speech versus the textual perfection of prepared speech.

[6] Or 'English': a good place to discuss 'British' versus 'English' and the often mistaken, but pseudo-standardized use of 'English' rather than 'British' in many countries.

[7] Again, students should get used to taking notes of figures and being accurate in translating them. Currency exchanges are good for both terminology and exchange rates (general economic cultural knowledge).

[8] We have signalled a change of turns here – allowing the interpreter to take her turn – because the change of topic (from using ready-made to fresh pasta), also signalled by a question mark, would in natural speech require an immediate reply. The trainer can, of course, choose to ignore this, but it might be disconcerting for the listener. For the interpreter, the lack of logical coherence could impede memory and rendition, which may seem unhelpful, but can on the other hand be used for memory practice.

For the appropriate length of chunks, the trainer will have to find the ideal balance between long chunks for memory practice and creating conversation turns that are as unobtrusive as possible, and to follow a natural flow of conversational turns with the interlocutor responding to each utterance where a response would usually be required. The trainer will also have to decide whether or not they want to practise 'realistic dialogic turn-taking' or consecutive interpreting practice with longer chunks.

[9] Reference is made here to Italian food habits. It is useful to include local traditions and discuss them in business dialogues.

[10] When doing food-dialogues, it can be fun for the students to discuss celebrity chefs like Jamie Oliver, and TV programmes like *The Naked Chef*, which they can watch to improve their comprehension skills as well cuisine terminology.

[11] Translating jokes is a challenge, even for experienced interpreters. Trainers may omit this sentence and the last one, but we think students should try and convey the message (also through non-verbal communication) even though the joke is not necessarily going to be funny in the target language. Trainers can discuss at the end of the lesson how to translate jokes and quips.

6.2   The health sector

Ideally, the main differences existing between English and Italian medical language and healthcare jargon should be discussed before embarking on these dialogues. Suggestions should be given on how to translate medical terms in both scientific and more informal registers. Cross-cultural issues should be covered too, for example by analysing issues that are considered taboo and the organization of the health systems in both L1 (first language) and L2 (second language) cultures.

Dialogue 6   A flu vaccination

SITUATION: An American citizen, aged 72, is staying at his daughter's home in Forlí, Italy. The GP has prescribed a vaccine.

IT = Italian speaker (the doctor)
E = English speaker (Mr Rudas, the American patient)
Level of difficulty: moderately difficult.
This dialogue contains several medical terms, and chunks are quite long. Moreover, there is a difference in register between the doctor's and the patient's language, and the latter is being quite argumentative; trainees may find this difficult. There is some underlying tension, due to a reluctant and rather grouchy patient.
Terms supplied: contagion (*contagio*), discomfort (*disagio*), debilitating disease (*patologia debilitante*), epidemic (*epidemia*), flu jab (UK), flu shot (US) or injection (*vaccino antinfluenzale*), GP (UK), family doctor (US) (*medico di base, medico di famiglia*), residential care home (*casa di riposo*). Other terms, like asthma, breathing problems, diabetes, heart disease, immune defence system, steroid treatment, can be discussed if they cause problems, but it can be a good idea to test them without any help from the trainer. [1]

IT: Buongiorno, signore. Come si chiama? *Good morning. What's Your name?*
E: … Rudas
IT: Mi deve cortesemente dire quanti anni ha. [2] *Could You please tell me how old You are?*
E: Seventy-two.
IT: Da quanto tempo si trova in Italia? *How long have You been living in Italy?*

E: I think four, five years. I came here after I retired.

IT: Bene. Quando è andato in pensione? *Good. When did You retire?*

E: Four years ago. When I retired I was 68. Now I'm staying at my daughter's, doctor. My daughter wanted me to come here, but I don't like doctors and I don't like hospitals. [3]

IT: Capisco. *I see.*

E: And anyway, I feel great. Why would I get the flu?

IT: Sì, vedo che lei è in ottime condizioni di salute e forse non ne comprende appieno la necessità, ma mi creda, sua figlia ha pienamente ragione. // Quando si arriva all'autunno, e noi siamo già all'inizio di novembre, è molto importare prevenire il virus influenzale che si presenterà fra poco, in inverno. Inoltre lei ha una bella tosse, e sua figlia mi ha detto che soffre di asma. *Weeell yes, I can see that You are perfectly fit and healthy now, and maybe You find it difficult to understand why You need a vaccination* [4], *but believe me, Your daughter is right. // In autumn – and it's already early November now – it's essential to prevent the influenza virus that will be coming soon, in winter. You also have a bad cough, and Your daughter told me You also suffer from asthma.*

E: I think she should mind her own business.[5]

IT: Se si hanno più di 65 anni, come nel suo caso, è molto importante affidarsi ad un vaccino. Inoltre se si ha un grave disturbo cardiaco o di tipo respiratorio, come l'asma, o il diabete, o si hanno delle basse difese immunitarie dopo una terapia con steroidi o una terapia oncologica [6], si ha bisogno della vaccinazione. // Lei credo che rientri in questo gruppo di pazienti solo per l'asma, vero? *If You are over 65 – as You are – getting a vaccination is important. Furthermore, if You have a severe heart disease or a breathing problem, like asthma, or diabetes, or You have a lowered immune defence system following steroid or cancer treatment, You need a vaccination. // I think You fall within this group of patients for the asthma alone, no?*

E: Hmm. I've been smoking for 45 years. I'm 72 now. Why would I need a shot?

IT: Non dimentichi che la vaccinazione è il mezzo più efficace e sicuro a nostra disposizione per prevenire l'influenza e per ridurre le complicanze, che possono essere rischiose soprattutto per le persone con patologie croniche. *You should not forget that vaccines are the most efficient and safest tools we have to prevent the flu and reduce the risk of complications that might be dangerous, especially for patients with chronic conditions.*[7]

E: But will a vaccine stop me from getting sick? I've always hated shots.[8]

IT: No, il vaccino la proteggerà solamente dall'influenza. Quindi potrà prendere degli altri virus, è vero, ma di solito l'influenza può causare dei seri problemi all'anziano, come le accennavo prima, quindi è meglio fare il vaccino. // Lei ha mai fatto un vaccino negli Stati Uniti? *No, this vaccine will only protect You from influenza. So You might still be subject to other forms of virus – You're right – but often flu can cause serious problems for elderly patients, as I was telling You earlier. That's why I suggest that You should get a vaccine. //* **[9]** *Did You ever get a vaccination in the United States?*

E: I don't know, don't remember. I don't put needles in my body, I don't do drugs. **[10]**

IT: Mi spiace contraddirla, ma non sono d'accordo. La vaccinazione antinfluenzale è considerata necessaria per tutte le persone (bambini e adulti) con condizioni di rischio per la salute, per le persone di età pari o superiore ai 65 anni, per il personale sanitario e anche per il personale che è a contatto con animali. *I'm sorry to contradict You, but I don't agree.* **[11]** *The flu jab is essential for everyone (children and the elderly) in conditions that carry health risks. We're talking about people aged 65 or over, healthcare professionals and also people who are in touch with animals in the workplace.*

E: I've never been sick a day of my life. And, I don't think it is safe.

IT: No, guardi, deve stare tranquillo, la vaccinazione contro l'influenza è sicura e efficace. Inoltre una febbre alta può essere molto debilitante per una persona della sua età, senza volerla naturalmente offendere, Sig. Rudas. *Honestly, You needn't worry* **[12]**, *flu jabs are safe and effective. Moreover, high fever can be debilitating for a person of Your age. I don't mean to be offensive, Mr Rudas.*

E: Well, I feel young and healthy. You know, when You were a baby, I was fighting in the war.

IT: Certo, vedo che lei sta bene, ma una febbre alta nell'anziano può comportare anche malattie quali la bronchite e la polmonite, che potrebbero costringerla a venire regolarmente in ospedale per curarsi. // Non la voglio assolutamente spaventare, Sig. Rudas, ma ci sono delle persone anziane che possono andare incontro ad un peggioramento consistente della salute solo per una brutta influenza nel periodo invernale, che li butta molto giù. L'influenza è una malattia altamente infettiva, che si diffonde molto rapidamente con gli starnuti e la tosse, e la trasmettono persone portatrici del virus. Lei ha dei nipoti, quindi potrebbe trasmettere loro dei virus. *Sure, I can see that You are well, but a high temperature in an elderly person can lead to bronchitis or pneumonia so they might be forced to go to hospital for treatment anyway. // I don't want to frighten You, Mr Rudas, but there are some elderly people whose state of health consistently deteriorates after a bad winter flu which leaves them very weak. The flu is a highly infectious disease that is spread very rapidly through sneezing and coughing, it's transmitted by carriers. You have grandchildren, and You might transmit the virus to them.*

E: But there's no need to have a flu jab now that it's sunny and it's not that cold any longer; why should I get a flu jab? But mmm, I was just wondering, when am I most at risk from this flu you keep talking about?

IT: Guardi Sig. Rudas, l'influenza compare ogni inverno, di solito per un periodo breve di qualche settimana, e di conseguenza molti si ammalano nello stesso periodo di tempo. In anni particolarmente sfortunati si può avere una vera e propria epidemia. E' quasi impossibile prevedere quanti casi ci saranno quest'anno. *Mr Rudas, the flu comes every winter, usually for a brief period, for a few weeks; consequently a lot of people get ill at the same time. In particularly bad years a real epidemic may occur. It's almost impossible to predict how many cases there will be this year.*

E: But if I had the jab last year, do I need it again now? I hear that flu viruses keep changing all the time …

IT: Ha ragione, Sig. Rudas, i virus cambiano di frequente, e questo significa che questo inverno l'influenza sarà diversa da quella dello scorso inverno, e anche il vaccino sarà diverso. *You're right, Mr Rudas, viruses change frequently, and this means that this winter the flu will be different from last year's flu, and the vaccine will also be different.*

E: How long is this jab going to protect me?

IT: Attualmente il vaccino protegge il paziente per circa un anno. *At present the vaccine protects patients for approximately one year.*

E: I get the flu from getting this shot?

IT: No, stia tranquillo Sig. Rudas, il vaccino non può causarle influenza. *No, don't worry, Mr Rudas, You won't get the flu from the vaccine.*

E: Are there any side-effects? [13]

IT: Beh, si deve sempre prevedere la possibilità di avere effetti collaterali. C'è chi ha qualche linea di febbre, o dolori muscolari per un paio di giorni dopo la vaccinazione, e potrebbe avere un po' di dolore nel braccio nel punto in cui si è effettuata l'iniezione. [14] Ogni altro tipo di reazione è molto rara. *Well, You cannot completely rule out any side-effects. Some people get either just a little bit of fever, or muscle pain, for a couple of days after the vaccine, or they may feel pain in their arm where they had the injection. Any other type of reaction is very rare.*

E: Am I 100% sure that I won't get the flu after the jab?

IT: Nessun vaccino è efficace al 100%, ma la maggior parte delle persone vaccinate non prenderà l'influenza. *No vaccine is 100% effective, but most people who get vaccinated will not get the flu.*

E: What if I get it?

IT: Se prenderà l'influenza, probabilmente sarà in una forma più lieve. *If You catch the flu, it'll probably be a milder form.*

E: Isn't it too late for the flu jab now? It's already November 10th ...

IT: No, Sig. Rudas, lei può sottoporsi alla vaccinazione in qualunque momento durante l'autunno e all'inizio dell'inverno. *No, Mr Rudas, You can get vaccinated at any time during the autumn and the beginning of winter.*

E: I had a mild allergy once, because I'd eaten too many eggs. Is this going to be a problem?

IT. Soltanto se si ha una forte allergia alle uova non è raccomandabile sottoporsi alla vaccinazione. Se si è trattato di un solo episodio, che non si è mai più ripetuto, non dovrebbe correre assolutamente alcun rischio. // Naturalmente possiamo metterci in contatto con il suo medico di famiglia, anche per e-mail, e sentire cosa ne pensa. *If You have a very strong allergy to eggs, then I wouldn't advise You to get a vaccination, but only in that case. If it was an isolated episode – that never occurred again – it is very unlikely that You will run any risk whatsoever. // Naturally we can contact your GP, even by e-mail, and ask for his or her opinion.*

E: That might be a good idea. If I decide to have the vaccination, can I come back here?

IT: Sì, certamente, se non riesco a convincerla di farlo adesso ... Può anche chiedere all'infermiera di venire a domicilio, a casa di sua figlia, per misurarle la pressione sanguigna. Spero che lei si deciderà e tornerà presto a trovarmi. Arrivederci, Sig. Rudas. *Yes of course, if I can't convince You to do it now ... You can also ask the nurse to come to Your daughter's home to measure Your blood pressure. // I hope You'll make up Your mind and You'll come to see me soon. Good-bye Mr Rudas.*

Commentary

[1] In this case it is essential to supply trainees with quite a long list of terms and words in context, and to differentiate British English and American English forms (*shot*, *jab* and *injection*). Relevant medical-related phrasal verbs and verb constructions ('to have a headache', 'to have/suffer from

a disease', 'to sustain an injury') and basic medical terminology such as *condition, illness, disease, ailment, vaccination, vaccine, immunization,* etc., could be introduced at this stage. Students are always encouraged to look for synonyms or similar expressions with a different register. This is good practice for vocabulary enhancement, paraphrasing, register practice, synonyms etc., so that if the first word does not come to mind, they immediately have another one ready.

[2] This politeness remark *Mi deve cortesemente dire* – literally 'You must kindly tell me' – can be discussed. The doctor is talking to an elderly man and he is trying to convince him to accept a flu vaccination. We are dealing with an argumentative patient here and this is the perfect opportunity to talk about the interpersonal – and not just cultural – aspects of doctor–patient communication. The patient is defensive and anxious and the doctor is applying both scientific (i.e. the connection between age and the flu) and interpersonal communication strategies to persuade him to accept his treatment plan.

[3] This dialogue, with the elderly American patient's insistence that he is perfectly well and doesn't need a vaccination, and the Italian doctor's cautious attitude, is also a good opportunity to talk about different ways of perceiving and expressing illness and pain cross-culturally: the Anglo-Saxon stoic approach versus the more Mediterranean tendency to complain more freely and verbalize illness more. (We make this point at the risk of propagating stereotypes.) Galanti's (2002) book provides a wealth of interesting cross-cultural examples from the health sector.

[4] The doctor uses quite a high register – as Italian doctors normally do – and trainees might want to adopt a lower register in L2.

[5] The elderly patient is quite impolite here, and the interpreter should not really get involved. One might discuss impartiality issues here as well as the need for interpreters not to take sides with either the patient or the service provider.

[6] The term *cancer* can be discussed and the evolution of the use of *cancer, tumour, oncological treatment,* etc. This entails cross-cultural issues of informed consent and the breaking of bad news, i.e. whether the patient should be told they suffer from cancer, and in what way the news should be broken. In Italy, the official policy towards the disclosure of information to patients has changed and clinicians are required to inform patients fully about their illness.

[7] Here again, trainees' attention is drawn to the different registers used by the doctor and the patient. (We have lowered the register in the translation into English, which we feel is more natural.)

[8] Another point of entry to discuss differences between British and American English. Students are expected to choose either British or American English lexis, orthography and pronunciation.

[9] It is a good idea to make breaks in chunks after a list or after the end of a long unit of thought/ reasoning etc. but before asking a final question,

even though that final question follows on from the previous utterance. Although it would seem natural to keep them together, it is very easy for the students to forget either the beginning or the end when the sentence form changes (from declarative to interrogative, for example).

[10] Notice the patient's colloquial and colourful expressions and the American use of '*drugs*'.

[11] Notice the politeness markers, even though the clinician is assertive in making his suggestions. This might also be an opportunity to discuss various expressions of agreement/disagreement and discuss the degrees of intensity and/or formality each of these convey ('I don't share your view', 'I'm not convinced', 'I don't agree with you', 'You're wrong there', 'No, you're wrong', etc.).

[12] Italians very often use the expression 'You don't have to worry' instead of 'Don't worry'. Modals used in the imperative format (*must, have to,* and *should, intend, mean to*) and indicating degrees of probability and ability (*may, might* and *can, be able to)* are also notoriously difficult for Italians but extremely important to convey in natural-sounding – and thus convincing – speech. An interpreter who produces natural-sounding idiomatic and confident renditions will also convince the listener more easily that the translation is accurate.

[13] In this case it is the patient that uses this medical term first and thus helps the student translate the following chunk by giving the L2 term.

[14] This is quite a long sentence which needs to be mentally simplified or summarized before translating, to avoid a heavy and awkward sentence structure.

Dialogue 7   High drama in the Seychelles

SITUATION: A group of Italian tourists are on vacation in the Seychelles. They are bathing at an isolated beach, far away from the resort they are staying at, on the other side of the island. One of the elderly ladies seems to have fainted while bathing and a young man – the only one who speaks English fluently – rescues her and brings her back to the shore. He translates for the woman and her husband in the following dialogue. In the first – and longest – part of the dialogue, he is speaking to the telephone operator and translating for one of the tourists who is performing artificial respiration. This dialogue is hardly representative of the type of situation interpreter trainees will find themselves in, but apart from the pedagogical value of a high density of technical but also non-complex medical terminology (also useful logistical terminology pertaining to touching and moving) we find that such 'dramatic' scenarios, especially when performed in the group-format, are conducive to class cohesion and motivation. By using an ad hoc interpreter here, not only in the emergency situation on the beach but also in hospital, it can also be used to reflect on the use of ad hoc interpreting in emergency situations as well as the need for trained interpreter staff in hospitals. Trainers might even choose to insert deliberate errors that would be plausible in such an ad hoc situation.

The situation has been made deliberately awkward logistically, with the interpreting taking place over the telephone. Using the telephone as a communication channel is highly artificial the way it has been used in this context, but it allows the trainers and students to reflect on the more practical and logistical aspects of interpreting if they so choose (moving around, acoustic interference, etc.). Although professional telephone interpreting is performed very differently, and nowadays often with video equipment, this could also be an opportunity to discuss remote interpreting in a more entertaining format. Issues such as the lack of non-verbal communication and physical proximity to the patient, tone of voice, emotions and fear evoked by emergency situations, and so on, can then be addressed. This dialogue is quite fun to perform and it requires a lot of 'acting'; there is a lot of moving around, gesturing and body language involved, which can be moderately confusing.

IT = Italian speakers (tourist, husband)
E = English speakers (telephone operator at the hospital, doctor)
Level of difficulty: difficult.

There might be some emotional stress involved which gives trainers the chance to discuss issues like interpreter support from psychologists or colleagues.

Terms supplied: artificial respiration (*ventilazione assistita*), airways (*vie aeree*), pulse (*polso*).

IT: (Young man) Adesso chiamo il 113 [1] dal mio telefonino, per fortuna avevo preso il numero ieri. *I'll call 113 from my cell phone. Luckily I noted down the number yesterday.* [This part does not need to be translated, as it is more of an introduction to the scene. From now on the young man acts as interpreter between the English-speaking tourists and the health professionals/ telephone operator. The first person singular should be used here as default interpreting format, even if the interpreter in the dialogue is not professional.]

IT: Ho una signora anziana di fronte a me, credo sia svenuta in mare, mentre faceva il bagno. E' molto pallida e fa fatica a respirare; sta soffocando. [2] *I'm with an old lady. I think she fainted while she was bathing. She is very pale and has breathing difficulties. She's suffocating.*

E: (Hospital operator) Now, don't panic. It's essential that no one panics. [3]

IT: Ma cosa devo fare io? *What should I do?*

E: Your friend has probably inhaled water into her lungs.

IT: Per favore mi dica cosa dobbiamo fare. Il viso è completamente bianco, ha le labbra blu e sento appena il polso. *Please tell me what to do. Her face is completely white, her lips are blue and I can barely feel her pulse.*

E: Don't worry. I'm going to help you and explain to you what you need to do. Unfortunately you're too far away from the nearest emergency department. I'll call an ambulance immediately, but it'll take some time. // Now, put your ear close to her mouth and tell me if you can hear or feel any breathing.

IT: Non sento alcun rumore, non sento neanche l'aria. *I can't hear anything; I can't even hear any air passing.*

E: Ok. Now this is what you must do. You must perform mouth-to-mouth resuscitation, artificial respiration. Follow my instructions carefully.

IT: La sento, sono pronto. *I'm listening; ready now.*

E: The first thing you must do is to make sure that she has a clear airway. It's crucial that her respiratory passage is not blocked by her tongue or by dentures. Can you do that? You need to tilt her head downwards.

IT: Sì. Le ho inclinato la testa verso il basso e controllato che la lingua non le blocchi la gola impedendole di respirare. E adesso, che cosa devo fare? *Yes, I've tilted her head downwards and I've checked that her tongue isn't blocking the throat and preventing her from breathing. And now what should I do?*

E: Now make sure that you eliminate any residual water from the air passage. Keep her head tilted downwards.

IT: Sì, lo sto facendo. Le esce un sacco d'acqua dalla bocca. Deve aver inghiottito un sacco di acqua di mare. Sono pronto. *Yes, ok, that's what I'm doing. There's lots of water coming out of her mouth. She must have swallowed a lot of sea water. What now?*

E: Now listen very carefully. Hold the lady's mouth closer and put your lips over her mouth and nose – both mouth and nose together – that's important. Then blow, not too hard, gently, but enough to provide her with oxygen. Then remove your lips. Can you see her chest rising?

IT: Sì, ho soffiato aria nella bocca e nel naso, non troppo forte, ma non si muove nulla, che cosa devo fare adesso? *Yes, I've blown air into her mouth and her nose, not too hard, but nothing is moving. What should I do now?*

E: Just continue slowly and gently. Repeat the procedure every 5 to 10 seconds or so. Wait till I say 'now'.

IT: Seguo le sue istruzioni. Respiro sulla bocca e sul naso. Sì, adesso il petto si alza e si abbassa mentre respiro nella bocca e nel naso e poi smetto, ma non respira da sola. *I'm following your instructions. I'm blowing air into her mouth and nose – now her chest is rising [4] and deflating while I'm breathing into her mouth and nose, and now I've stopped, but she's not breathing on her own.*

E: The patient's chest should rise when you breathe into her nose and deflate when you let go – with expiration. That's what happens. Just keep breathing regularly. // Pause for 7 to 8 seconds, then start again. The alternate compression and relaxation of the chest wall will stimulate the blood supply to the heart and provide oxygen for the lungs. [5]

IT: Sì, adesso il viso è ancora pallido e le labbra sono blu. Immagino che questo significhi che c'è ancora carenza di ossigeno, ma sembra ci sia un debole polso. *Yes, her face is still pale and her lips are still blue. This probably means that there's still a shortage of oxygen, but I can feel her pulse faintly.*

E: Ok, now that the most important thing has been taken care of, you must make sure that you maintain your friend's body heat with blankets [6]. Can you do that?

IT: Abbiamo degli asciugami qui con noi, possono andar bene? *We have some towels here, can we use them?*

E: As long as they're dry, but whatever you do don't cover her with wet towels.

IT: No, penso che così le farei perdere ancor più calore corporeo. Per quanto tempo devo andare avanti? *No, then she'd lose even more body heat, I suppose. How long shall I go on for?*

E: Keep going till the ambulance arrives. A person can be kept alive artificially for as long as an hour, so don't lose heart.

IT: Mi sembra di essere già andato avanti per ore, ma probabilmente sono passati solo dieci minuti. Devo continuare con la ventilazione assistita una volta che è arrivata l'ambulanza? *It feels like I've been doing this for hours, but I probably only started ten minutes ago. Shall I continue with the artificial respiration once the ambulance has arrived?*

E: The ambulance will have mechanical methods for artificial respiration which are more effective – and then once she gets to the hospital they can use electric shock treatment as well as pharmacological treatment.

IT: Finalmente, ecco, sento che sta arrivando l'ambulanza. Vado avanti finché non arrivano. *At last! I can hear the ambulance. I'll keep going until they arrive.*

E: Good, I'll leave you in their hands now. I wish you all the best.

IT: Grazie di tutto. Non so cosa avremmo fatto senza di lei … *Thank you so much. I don't know what we would have done without you …*

E: Only doing my job.

[In the hospital some hours later: The doctor is examining the patient's X-rays and clinical record and explaining them to the patient's husband. A friend is acting as the interpreter between the doctor and the patient's husband.]

IT: Dottore, grazie di tutto. Mi può dare notizie sulla salute di mia moglie? *Doctor, thanks so much. Can you give me any information about my wife's health condition?*

E: (Husband)   She seems much better now. Of course she needs to be taken care of. She'll stay in hospital for a couple of days, until she recovers completely. We'll need to carry out some tests to check her general state of health.

IT: Mi può dire di cosa si è trattato esattamente? *Can you please tell me what happened exactly?*

E:　Now, it seems to me that your wife may have suffered from heat stroke while still in the water – there was no sign of muscular tension – and as a result she fainted, causing water inhalation in the lungs and subsequent suffocation – in other words, she was drowning.

IT:　C'è mancato poco, se quel ragazzo non si accorgeva subito ... Pensi che stavo prendendo il sole, non mi sarei accorto di nulla! Mi sento così in colpa. *She nearly died, you know, if that boy hadn't noticed that she was in trouble ... I was sunbathing, you know, I wouldn't even have noticed! I feel I'm the one to blame.*

E:　It must have been a great shock. The lack of blood supply and oxygen to the lungs, and therefore to the brain during the 9 or 10 minutes that passed between her passing out and the application of artificial respiration, may have led to brain damage, which is what we are trying to ascertain through the brain scan we've just performed. This hospital, as you can see, is equipped with state-of-the-art technology and, being a small hospital, we have access to our machines virtually round the clock.

IT:　Dottore, ringrazio di cuore lei e i suoi colleghi e rimango in attesa di ulteriore notizie. Grazie. *Doctor, I really want to thank you and your colleagues. I'll be waiting for more information. Thank you so much.*

Commentary

[1] Students should know the emergency numbers in the relevant countries, 999 in many English-speaking countries, for example. We have used 113 here, the Italian emergency number, but this of course should be adapted to L1 and L2 circumstances.

[2] Differences in register should be highlighted in this dialogue. Usually, but especially in emergency situations like this one, the main interpreting task is that of conveying the message, and if the student is not familiar with a medical term in the source language, she should explain the term using simple words, or even gestures. Here the Italian rescuer is using a very colloquial term that can be translated using a higher register if the interpreter is talking to a physician. However, in an emergency situation, accuracy and clarity are far more important than stylistic features. Indeed, register is not the main concern in this kind of situation, but as it is a simulation, students should be made aware of these issues.

[3] This emergency situation allows us to discuss the role of the interpreter – who should remain calm and detached.

[4] Funnily enough, seemingly simple expressions indicating body movement, both involuntary and voluntary, are often difficult for students; one might stress the need to explain the action to be performed, even though students are not familiar with the exact translation.

[5] These passages are dense with technical terminology and good for practice, but might have to be read slowly or repeated a few times. This might not be appropriate for an emergency situation, but it is good for terminology practice.

[6] Even a simple word like this (*blanket*) can prove to be difficult when trainees are tense, also because of the emergency situation, even if it is only simulated. Stress the importance of making themselves understood, even if they use a different word.

Dialogue 8   A delivery in Australia

SITUATION: We are in an Australian hospital. A young Italian woman is at the emergency ward because of early labour pains. A nurse, who uses a high technical register, is filling in a form and has just asked her name. The hospital has an in-house interpreting service and interpreters are available 24 hours a day.

IT = Italian speaker (patient)
E = English speakers (nurse, midwife)
Level of difficulty: difficult.

Introductory note: The dialogue contains a long list of specific gynaecological terminology, as well as general medical terminology, although the register is quite colloquial. Students may find it difficult to respond promptly to reformulating difficulties when interpreting into English. This emergency situation also contains many colloquial expressions of pain and annoyance and a great deal of moving around, which can be 'acted out' or simply referred to. It combines colloquial expressions with technical terms and explanations of how the body works (delivering a baby).

Terms supplied: pregnancy (*gravidanza*), miscarriage (*aborto spontaneo*), abortion (*aborto*), labour pains (*doglie*), umbilical cord (*cordone ombelicale*), Caesarean section or C-section (*taglio cesareo*), anaesthesia (*anestesia*).

IT: Mi chiamo Elena Biagi. Mi aiuti per favore, mi sento molto male, ho le doglie, adesso – oh che male! *My name is Elena Biagi. Please help me, I'm in such pain, I have labour pains – ow it hurts!* [1]

E: (Nurse) Hm... Hang on a moment please, I'll just have to dig up your clinical record [2] – oh here it is! You've been attending your local health clinic during the prenatal period, I see. And you're a week before your due date. How long have you been feeling pain [3]?

IT: I dolori sono cominciati un'ora fa – almeno un'ora fa. *The pain started an hour ago, at least an hour ago.*

E: Well, that means we have plenty of time. Just breathe deeply. Now, did you have any problems during pregnancy? Is this your first pregnancy?

IT: Nessun problema grave, no, ho solo la pressione bassa. Ho avuto un aborto spontaneo e questa volta mi hanno tenuto sotto controllo i primi mesi. //Ah – vedo che mi vuole misurare la pressione, vero? E deve fare anche un prelievo? *I've never had any major problems. I've just got low blood pressure. I've had a miscarriage and this time I was followed very closely during the first months of the pregnancy. // I see You want to take my blood pressure? And a blood sample as well?* [4]

E: Yes, we'll need that just in case of complications. Have you thought about what form of anaesthesia you'd like to use – or perhaps you'd prefer not to use any?

IT: Speravo di poter fare l'epidurale, ma se lei mi consiglia un altro tipo di anestesia, faccio quello che mi dice lei. *I was hoping I might have an epidural, but if You could suggest any other type of anaesthesia I'll do what You tell me to do.*

E: I'm afraid an epidural has to be administered three hours before the delivery, so I can't guarantee that we'll be in time. Here at the hospital we often use 'gas and air'[5], a form of gas that relieves much of the pain – you inhale it when the pain becomes too intense. // Alternatively, we have morphine injections injected in the tissue near the birth canal in minimal dosages, but very effective. We sometimes administer the analgesic in drips, but I don't think that will be necessary in your case.

IT: Vorrei fare l'anestesia per inalazione – ahi, che dolore, adesso le contrazioni mi stanno venendo più spesso; ma guarda – tutto questo bagnato? *I'd like to have 'gas and air'– oh, it really hurts; the contractions are coming more often now, but look, I'm all wet now…*

E: Oh not to worry, that'll be the membranes rupturing, or 'the waters breaking' as we say. [6]

IT: Cosa sta succedendo esattamente, mi puoi spiegare? *What is happening? Could you please tell me what's going on?*

E: What is happening is that the water – the amniotic fluid – in the womb is pressing downwards and the pressure has broken the membrane of the amniotic sac, allowing the fluid to leak out. // That means that the foetus – the baby – is almost ready to come out and is beginning to push, or rather be pushed, helped by your contractions. // Try to breathe regularly and slowly and push when I give the word. [7]

IT: Va bene, farò del mio meglio. *All right, I'll do my best…*

E: What happens now is that we'll take you to a more comfortable room where the anaesthetist will set the analgesia dosage and a midwife – she'll be along any minute now – will come by regularly to check your dilation – that is the opening of the womb – until you're ready to go into the delivery room. // The obstetrician will come to see you when you go in and we'll be with you during the actual delivery. // Here, do you feel strong enough to walk? Take the interpreter's arm, that should do the trick [8].

IT: Ecco, così riesco a camminare. // Ascolta infermiera, mia sorella diceva che quando ha partorito lei, il cordone ombelicale era annodato attorno alla testa della bambina ed era tutta blu, sembrava stesse per soffocare – non succederà anche a me, vero? *Oh, yes, this way I can walk. // Nurse, my sister told me that when she delivered her baby the umbilical cord was wrapped around the neck of the baby – she was all blue, and it looked like she was about to choke. That isn't going to happen to me too, is it?* [9]

E: Let's hope not, but it's actually not that rare. It sounds frightening, but the midwife just slips a finger under the cord to release it, it isn't difficult. Right, come along then – ah, here we are, here's your room – get undressed and slip on this tunic; make yourself comfortable here on the bed. // Ah here's the midwife – what, 7 centimetres already? Only 3 to go. This is by far the most painful part of the process, so stiff upper lip [10] – and good luck! I'll hand you over to the midwife now.

IT: Ho dei crampi insopportabili adesso, dammi un po' di anestesia, per favore. *The cramps are unbearable now. Please give me some pain relief.*

E: (Midwife) Here, hold this mask over your nose and breathe slowly, that's right. // I have a suspicion that this might be an unexpected breech presentation [11] – I can't feel the baby's head and it feels like it's turned feet down.

IT: Cosa vuol dire parto podalico? Non riesci a sentire la testa del bambino? C'è qualcosa che non va? Cosa si puoi fare? Può vedere subito se va tutto bene, per favore? Dove mi state portando? Devo fare un taglio cesareo? E l'anestesia? Almeno adesso facciamo l'epidurale? *What do you mean by breech presentation? Can't you feel the baby's head? Is something wrong? What are we going to do? Can you check to make sure everything's all right? Where are you taking me? Am I going to have a Caesarean section? What about anaesthesia? Can I have an epidural now?* [12]

E: Into the delivery room immediately, help me, midwife – no, Alessandra, no interpreters beyond this point; I'm afraid the young lady will just have to try to communicate as best she can. Thank you for your help.

Commentary

[1] Both the patient and the nurse use colloquial medical forms in these passages. Sometimes this turns out to be more difficult for the students than more formal scientific language, where the Latin or Greek origin of the word may be helpful.

[2] Hospital terminology can be discussed here: *'to be admitted to hospital'*, *'clinical record'*, *'hospitalization'*, *'to be discharged'*, etc.

[3] Pain-related terminology can be discussed: *labour pains, painkillers*, etc. This topic also gives trainers the chance to discuss cross-cultural issues, such as how people of different cultures have different ways of showing their pain, or grief, or have different pain thresholds. (Italians are said to verbalize their pain very loudly, for example!)

[4] The patient is very nervous and keeps asking questions. After the student's rendition trainers can discuss whether the interpreter should be equally insistent. The interpreter should try to adopt an emotionally detached attitude even if she chooses to 'act out' the more overtly dramatic communication features.

[5] Many specific medical terms are encountered in this dialogue; some of them are discussed at the beginning of the lesson, but students can practise explaining the meaning of the word – after asking the interlocutor for clarification in the source language. In a real-life situation they may come across words they are not familiar with and might have to ask for an explanation and clarification. Also, various medical registers could be discussed here, e.g. *'anaesthesia'* versus *'painkiller'* or *'pain relief'*.

[6] This sentence allows trainees to practise both the scientific term and the colloquial form. Synonyms are also important. It might be a good idea to use several expressions with the same, or roughly the same, meaning when it really is important to make oneself understood, as in this case. It is good interpreting practice to use synonyms, and for the students to list them in their glossaries.

Here the patient changes from the polite 'You' form to the informal 'you' form when the pain becomes too intense to maintain the formal register.

[7] This long explanation clearly demonstrates that it is much easier for the interpreter to perform if she is familiar with the terminology.

[8] Inserting colloquial expressions in your dialogues normally calls students' attention to interesting solutions and allows them to enlarge their English vocabulary.

[9] In this passage too, the patient is very anxious. Trainers can stress how important it is for the interpreter to control her nerves and not get too involved when interpreting.

[10] This is an expression that students might not be familiar with. Culturally determined expressions like these can lead to stimulating discussions on differences between L1 and L2 cultures.

[11] Again, this is a very specific medical term that most students are not familiar with. Students are not told this term beforehand, and must try to infer the meaning from the context or ask for an explanation. To familiarize themselves with these topics, students can read scientific articles, or listen to interviews on the internet with patients, or with physicians talking about their diagnoses.

[12]   The patient is beginning to panic and the trainers and students will have to be creative while interpreting this passage and coordinate the chunks and emotions effectively (which can be great fun for role play). Here we find a string of desperate questions asked by the patient, not giving any time for the midwife to reply. We have not signalled any turn changes here, which effectively puts the student in great difficulty, showing how challenging these emergency situations can be. Trainers can of course adapt the turns to meet their specific needs.

Furthermore, this question shows that very often patients do not understand what healthcare professionals tell them and they need explanations. It is important to specify here that the interpreter is not expected to give them explanations, but simply to translate the patient's questions or to report the problem to service providers.

Dialogue 9   Interpreting at the Casualty Department

SITUATION: An English patient arrives at the Casualty Department in an Italian town with his wife [1].

IT = Italian speaker (Dr Moretti)
E = English speaker (Jane Rowlands, the patient's wife)
Level of difficulty: quite easy, though the dialogue contains several cardiological terms.
Terms supplied: chest pain (*dolore precordiale*), painkiller (*analgesia*), infarction (*infarto*), coronary angiography (*angiografia coronarica*), angioplasty (*angioplastica*), contrast medium (*mezzo di contrasto*)

IT: Buongiorno, Signora. *Good morning.*
IT: Sono il Dr. Moretti, Cardiologo. Allora vorrei sapere cosa è successo. *I'm Dr Moretti, I'm a cardiologist. I'd like to know what happened.*
E: We were sitting on the beach when John suddenly started clutching his chest [Mrs Rowlands demonstrates this movement] and seemed to be in great pain.
IT: Era un dolore precordiale, vero? Era molto forte? *It was a chest pain, wasn't it? Was it very intense?* [2]
E: Well, I don't know, that's what it looked like. It looked like he was in a lot of pain, so we called the ambulance. We were incredibly lucky because there was a physician sitting next to us on the beach, and he immediately phoned Your hospital on his mobile phone. The ambulance came very quickly.
IT: Lo faccia sedere qui. Lo tranquillizzi, è tutto sotto controllo. *Have him lie down here, and tell him not to worry. Everything is under control.*
E: At the beginning my husband thought this pain was due to his rheumatism. I called his General Practitioner in England on our way here and Dr Abbots told me that it might be rheumatism, and suggested that he should take a painkiller.
IT: Il dolore non è passato però, credo. Ecco, ora faremo un ECG (elettrocardiogramma), che ci permetterà di capire cosa è successo. // [Dopo 10 minuti]. Sì, l'ECG ha confermato che c'è un infarto in atto. *He's still feeling pain, I believe. Well, we are going to run an ECG, an electrocardiogram* [3], *to try to find out what happened.* // [After 10 minutes] *Yes, the ECG has confirmed that he's having an infarction.*
E: Oh my goodness, You're frightening me; is this very dangerous? What are You going to do now, Dr Moretti?

IT: Stia tranquilla signora, andrà tutto bene. // Ora abbiamo bisogno del consenso informato del Sig. Rowlands, ovvero della sua autorizzazione ad eseguire un'angiografia coronarica e poi un'angioplastica. *Don't worry, Mrs Rowlands, he'll be fine.* **[4]** // *We now need Mr Rowlands' informed consent, that is we need his authorization to perform a coronary angiography first and an angioplasty later.*

E: Yes, please John sign here [turning to the patient]. Thank you Dr Moretti.**[5]**

IT: Signora Rowlands, suo marito verrà ricoverato, e resterà in ospedale una decina di giorni, per permetterci di effettuare tutti gli esami che risulteranno necessari ed escludere ogni complicanza. Poi verrà dimesso se non ci sarà alcuna complicanza. // Purtroppo non abbiamo un servizio interpreti durante la notte, anche se l'abbiamo già richiesto da tempo all'Ufficio Amministrazione. Fortunatamente alle ore 20.00 monterà in servizio un collega, il Dr. Remoli, che parla bene l'inglese. *Mrs Rowlands, Your husband will be hospitalized now* **[6]***, and he will stay in hospital for ten days, approximately, so that we can run all the necessary tests and exclude any complications. Then, if no complications arise, he'll be discharged from hospital.* // *Unfortunately our hospital doesn't have an interpreting service during the night, even though we sent in a formal request to our Managing Division long ago. Luckily, Dr Remoli – who speaks English fluently – will be on duty at 8 this evening.*

IT: Ho ancora una domanda per lei: suo marito ha mai avuto problemi di allergia? E' una domanda che devo porle, perché il paziente potrebbe essere allergico al mezzo di contrasto. *I have one last question for You: has Your husband ever had any allergic reactions? I have to ask You this question because the patient may be allergic to the contrast medium.*

E: John never had problems with allergies that I know of.

IT: Nel caso vi fossero problemi passeremmo ad una terapia anti-allergica con cortisone. Ecco, ora l'infermiera le accompagnerà alla Stanza N. 2. *Should there be any problems, we'll switch to a cortisone-based anti-allergy treatment. Now the nurse will take You to room no. 2.*

Commentary

[1] Both British English and American expressions are examined: Casualty Department, A&E (Accident and Emergency), ER (Emergency Room), etc.

[2] This expression allows the trainers to expand on the concept of pain and the related terms: *acute, mild, sharp, throbbing pain, painkiller,* and expressions such as 'does your (ankle) hurt?', etc.

[3] Acronyms always pose problems and should be explained.

[4] This is an opportunity to discuss the doctor's claim that 'he'll be fine' to reassure the patient's wife. Time permitting, it would be interesting to discuss the issue of healthcare interpreters' professional liability.

[5] In this case the patient's relative trusts the doctor. It's a good idea to bring in examples of more complicated cases where the issue of informed consent might not be so straightforward; for example, if the patient or the patient's relatives disagree, or the patient's family prefer not to give the patient too much information about the illness.

[6] Trainers can expand on related expressions: 'to be admitted to hospital', 'to be discharged from hospital', 'hospital stay', 'hospitalization' etc.

Dialogue 10   Interpreting for an osteoporosis awareness campaign

SITUATION: Mr Nelson, an American, is invited by the Osteoporosis Prevention Centre of Forlì to be their ambassador for an osteoporosis awareness campaign.

Introductory note: Being an interview, this dialogue, like Dialogue 5, also presents some interesting challenges. Note how the English speaker immediately starts to tell the interviewer about the inadequacies of his GP and is clearly very upset and angry when he re-lives these experiences through his narrative. This is useful to play out in class, both for the slightly confused formulation of his utterances (very natural given the situation) and for the opportunity to see how the students handle emotional language and body language. (We have used italics and exclamation marks to suggest emotional emphasis.)

This dialogue can be used to talk about the differences between patient–doctor communication in the consulting room or in an emergency situation and the orderly, turn-taking and progression of an interview format: registers, lexical density, greeting and leave-taking, politeness strategies, topic-changing, entry points, turn-taking and emotivity, and so on.

IT = Italian speaker (journalist)
E = English speaker (Mr Nelson)
Level of difficulty: difficult.
Very specific medical terminology is used, and the dialogue also contains other difficulties, such as translating units of measurement.
Terms supplied: osteoporosis (*osteoporosi*); DXA scan (*la metodica DEXA*); femur/ thighbone (*femore*); hip (*anca*); osteoporotic bone (*osso osteoporotico*) [1]. Several medical terms have not been supplied – to encourage the students to infer the meaning from the context. This will increase their self-confidence and train them to ask for explanations when they get stuck on a word.

IT: Buongiorno, Sig. Nelson, e grazie di aver accettato il nostro invito e di essere qui con noi oggi, al Centro per la Prevenzione dell'Osteoporosi di Forlì, a raccontarci la sua esperienza. Come ha saputo di essere affetto da osteoporosi? *Good morning, Mr Nelson. Thank You for accepting our invitation [2] and being here with us today at the Osteoporosis Prevention Centre [3] in Forlì, to tell us about Your experience. How did You learn You had osteoporosis? [4]*

E: Four years ago, when I was just 37, I began noticing a loss of height, which had been accelerating in the previous 12 months. I discovered that I had shrunk by two inches! Something the doctors told me was *impossible* for someone my age. // They told me it was just my *imagination* that I had gone from 5 feet 8 inches to 5 feet 6 inches. [5]

IT: Credo che non fosse la sua immaginazione, vero? *But it wasn't Your imagination, was it?*

E: No, indeed! It was osteoporosis. But the doctors didn't recognize it. It was only when I broke three ribs after walking into a door that it 'clicked'. // Then they began to realize that the fractures, my height loss and painful joints might be caused by osteoporosis. It certainly took them long enough...

IT: A quel punto è stato mandato da uno specialista? *Were You referred to a specialist then?* [6]

E: Actually, it was at *my* suggestion that my GP *finally* contacted a specialist, although it took almost a year to convince him. I can tell You I was pretty desperate; nothing else would explain my problems. [7]

IT: Probabilmente il medico non pensava che lei potesse avere l'osteoporosi, malattia che di solito colpisce le donne. *The physician probably didn't think You could have osteoporosis, as this disease usually affects women.*

E: Yes, exactly, that's the whole point. Osteoporosis does not usually affect men. I think the doctor was as surprised as I was when the DXA scan indicated osteoporosis.

IT: Cosa ci può dire dei fattori di rischio? *What do you know about the risk factors?*

E: Well, I tell You, I think the doctors should have recognized that height loss in a young man is not normal. And I had two years of chemotherapy for Hodgkins disease, as well as five years of inhibitors [8], which might have affected my disease too.

IT: Pensa quindi che sia necessario cambiare la lista dei fattori di rischio per l'osteoporosi? *Do You think it is necessary to change the risk factor list for osteoporosis?*

E: Yes, perhaps we should modify the risk factors reported to doctors and radiologists to include medical treatments involving cytotoxic compounds and antiviral drugs.

IT: Forse il medico di base dovrebbe anche essere più informato e quindi propenso a considerare l'osteoporosi un possibile problema anche per l'uomo, a qualsiasi età. *Maybe general practitioners [9] should be better informed and more receptive to the possibility that osteoporosis could affect men too, at any age.*

E: Yes, I agree. The good thing is that I now have a reason for my unexplained problems.

IT: Quali sono stati i problemi più difficili che ha dovuto affrontare? *What were the most difficult problems You had to tackle?*

E: The biggest embarrassment was that in order to get a bone density scan I had to attend a 'Well-Woman' clinic and endure some rather hostile 'get out of our women-only space' remarks from the other clients. // This was not the first time, as I also had to have a mammogram (which is an even more difficult and uncomfortable procedure on a man than a woman), since my disease caused a breast-enlarging lump to appear (which after a biopsy was found to be benign). // All of this seemed to defy the myth that both osteoporosis and breast cancer are women-only problems and age-related issues.

IT: Sono queste difficoltà che l'hanno spinta a partecipare attivamente alla campagna di prevenzione dell'osteoporosi negli Stati Uniti? *Are these the main difficulties that prompted You to become an ambassador for the osteoporosis prevention campaign in the United States?* [10]

E: I'd say yes. I am campaigning locally for recognition of these problems for both younger and non-female patients and also to establish more general clinics besides the 'Well-Woman' clinic. // Given the intimidating atmosphere, I think that many men in my situation would not have undergone the necessary tests. I hope my story will help other people in 'non-risk' categories to get a faster diagnosis.

IT: Vuole aggiungere qualcosa sugli esami a cui ha dovuto sottoporsi? *Is there anything else You would like to say about the tests You had to have?*

E: Sure. I had to undergo several DXA scans. I am relieved that my latest DXA scan shows marked improvement, following treatment with bisphosphonates and a high calcium and magnesium diet. I am now near the fracture threshold rather than well outside it. [11]

IT: Ha ancora dei problemi, dopo tutte le cure che ha seguito? *Do You still have problems, after all the treatments?*

E: The height loss [12] continues and I still have painful joints from osteoarthritis, partly because of the late diagnosis of my osteoporosis. I do things more carefully now and use a stick when walking, since I don't want to fall.

IT: Ha subito altre fratture? *Have You had any other fractures?*

E: While I haven't suffered any major fractures I still manage to break toes with monotonous regularity, but, perhaps fortunately, I also suffer from peripheral neuropathy so that after the initial pain I almost forget I've broken a bone.

IT: La prego di raccontarci della sua prima visita al Centro in cui le è stata diagnosticata l'osteopatia. *Please tell us about Your first visit to the Centre where You were diagnosed as having osteopathy.* **[13]**

E: When I was first referred to the 'Well-Woman Centre' I was given a very frosty reception. Now the clinic recognizes it has a more varied role and has become an independent osteoporosis and osteopathy centre with a much more open attitude. // Could this be due to the increasing number of male patients, I wonder?

IT: La ringraziamo molto, Sig. Nelson, di aver cortesemente risposto a tutte le nostre domande, grazie. *We thank You very much, Mr Nelson, for answering all our questions, thank You.*

Commentary

[1] We supplied the most technical and seemingly difficult words here, but found that many students, unexpectedly, got stuck on the more everyday words, like *shrink, ribs, joints, fracture* and *inch*.

[2] Trainers can explore idiomatic expressions to thank invited speakers and give them the floor. This occasion is not formal, and only colloquial expressions emerge, but suggestions for formal settings can be made as well.

[3] The name of the Centre allows trainers to examine the translation of names of organizations and the relevant acronyms. Some acronyms like WHO (OMS) have equivalents in the student's mother tongue, others require translation, like the Italian Azienda Sanitaria Locale ('Local Health Trust').

[4] This expression gives trainers the chance to discuss differences between English and Italian medical terms and the various registers used – 'essere affetto da osteoporosi' is much more formal than 'to have osteoporosis'.

[5] Converting units of measurement can be challenging. Students can prepare a small table with the most common measurements in L2.

[6] Terms like 'referral', 'to be referred to a specialist', can be examined here.

[7] The speaker is clearly both upset and angry. The trainers could try different tones here: excited, angry, sad, scandalized, etc. The trainers should also be careful to articulate carefully and slowly (despite the emotional tone), otherwise it can be very hard for the students to understand such emotional and slightly incoherent speech.

[8] Such specific medical terms can be difficult (*'inhibitors', 'risk factors', 'cytoxic compunds', 'antiviral drugs', 'DXA scan'* and *'biosulphates'*). They can be examined in further detail when analysing the students' renditions. Trainers are not doctors, however, and students can carry out additional specific terminological research at home.

[9]   Changing the singular (Italian version) into a plural (in English) as it is a
      general statement, is often a problem for students. Dialogues are always
      good for discussing points of grammar.

[10] Students could look for similar prevention campaigns on the internet and
      report on them in the following lesson.

[11] This part of the dialogue is rich in medical terms. As it is Dialogue 10, stu-
      dents should not be supplied with too many terms in advance, but should
      try and solve these terminological problems independently. Trainers can
      discuss them at the end of their rendition.

[12] Unexpectedly, this particular syntactic and lexical combination proved to
      be difficult for many Italian students; changing the word order and repeat-
      ing 'loss of height' helps if one sees that the student is puzzled.

[13] Here trainers could discuss how to translate the main suffixes found in
      medicine, like *-pathy*, and the relevant suffix in their mother tongue.

6.3 The legal sector

Dialogue 11 Applying for a residence permit

SITUATION: An American student is applying for a residence permit at the local police station and the immigration officer is giving him information on how to proceed.

Note: Terminological and cultural issues: target terminology tailored to the target culture and/or system

When interpreting in a legal setting there is often no clear equivalence in the source and target languages; for instance here when talking about financial solvency, if the migrant asks the police officer what amount of money he's referring to, the expression 'una somma equivalente all'assegno sociale' might be used in Italian, namely a welfare check (Am. English). In this case the interpreter is taking on the role of *message clarifier*, as she'll have to explain that according to the Italian welfare system this sum currently equals 409.65 euros and that people over 65 who are resident in Italy are eligible to apply for it.

Students were also very interested in the different ways the Italian *Codice Fiscale* might be translated into English, according to the country of origin of the migrant/client. As homework, they were asked to find more information on this term as well as on *Assegno Sociale* or welfare check/cheque. We then discussed various different versions: TFN (Tax File Number) in Australia, SIN (Social Insurance Number) in Canada, NIN (National Insurance Number) in England, PPS. (Personal Public Service) in Ireland, SSN (Social Security Number) in the USA. Here again, we stressed how important it is for an interpreter to prepare glossaries and to look for information in the literature and on the web.

IT = Italian speaker (immigration officer)
E = English speaker (applicant)
Level of difficulty: easy.
Only a few legal terms are used, while several bureaucratic differences emerge between the host country and the country of origin of the migrant, especially the social security systems. Explanations will probably be necessary.
Terms supplied: residence permit (*permesso di soggiorno*)

IT: Si? Come posso aiutarla? *Yes, how may I help You?*
E: I would like to have some information regarding the renewal of my residence permit.

IT: Può essere più preciso? *Can You please be more specific, Sir?* [1]
E: I've been here for one year and my permit expires next month. I would like to change my student permit into a work permit.
IT: Che tipo di lavoro ha trovato? Sa che per cambiare il suo attuale permesso in un permesso di lavoro deve avere un lavoro a tempo pieno con un datore di lavoro che paga tutte le tasse previste dalla legge italiana? *What type of job have You found? Do You know that in order to change Your existing permit into a work permit You must have a full-time job with an employer who will pay all the taxes required by Italian law?* [2]
E: Yes I do. My boss has filled it out and sent in all the necessary paperwork regarding my employment with his firm.
IT: Ok, quello che lei deve fare è riempire questi documenti in tre copie e poi tornare con una marca da bollo da 14,72 euro, quattro foto recenti, una dichiarazione di solvibilità, la conferma del suo padrone di casa che ha un posto in cui vivere e l'assicurazione sanitaria che la copre in caso di incidente. Le ho allegato un modulo che elenca tutti i documenti richiesti. *Ok, what You need to do is complete these documents in triplicate and then come back with a tax stamp for the value of 14.72 euros* [3], *four recent photographs, proof of financial solvency, an acknowledgement from Your landlord that You have a place to live, and health insurance that covers You in the event of an accident. I have attached a form that states all the requirements.*
E: I'm sorry, what do You mean by proof of financial solvency?
IT: Intendo una dichiarazione ufficiale dalla banca che dichiara che lei ha abbastanza fondi per sostenersi per tutto il periodo che rimane. *I mean an official bank statement saying that You have enough funds to cover You for the duration of Your stay.*
E: Will You give me all the necessary forms?
IT: Posso solo darle quello per il rinnovo e i moduli per il suo padrone di casa. Gli altri deve procurarseli dalle autorità in questione. *I can only provide You with the renewal form and the one that Your landlord has to fill in. The rest You will need to get from the relevant authorities.*
E: Ok, thank You.
IT: E deve portare il suo codice fiscale. Ne ha uno? *And You will need to bring Your national insurance number.*[4] *Do You have one?*
E: No, why? Do I need one?
IT: Si, per poter lavorare deve avere un codice fiscale. *Yes, You must have a national insurance number to be able to work.*
E: Where do I get one of those?

IT: Deve andare all'Ufficio delle Entrate, sa, un ufficio che rilascia il codice fiscale e si occupa di redditi e imposte. Lo può trovare nell'elenco telefonico. *You must go to the Public Revenue Office, You know, the office that issues tax codes and regulates incomes and taxes. You can find Your local office in the phone book.*

E: Ok, I think I have all the information I need to get started. Thank You.

IT: Prego. Buona fortuna. *You're welcome and good luck.* [5]

Commentary

[1] We have used 'Sir' here to suggest a polite register, although it may seem unusual to many English speakers. It is, however, a good opportunity to discuss forms of address and degree of formality, context etc., also between US and British English.

[2] This situation shows students how important it is for the interpreter to be familiar with the various procedures regulating immigration. Knowing the basics of the national healthcare and legal systems in both languages is essential. Students cannot become medical or legal experts of course, but touching upon these issues – through simulations and a few explanations – would suffice at this level of training. In this context, it can be very useful to maintain key bureaucratic terms, names, acronyms, institutions, etc in the original, as well as translating them, because it is in the original form rather than in the translation that the foreign speaker will be identifying and using the terms (e.g. *permesso di soggiorno* or *codice fiscale*). Indeed, many foreigners quickly learn the most essential terms, acronyms and institutions in the host country language that pertain to such issues as applying for a job, house-hunting, social service benefits, health services, bank accounts and bank loans, and so forth.

[3] Again, figures, currencies, and lists of documents are mnemonically challenging.

   This is also an excellent passage to discuss the level of bureaucracy in different countries and more specifically, bureaucratic language and register. The chunk is very long and dense, and can be adapted by the trainers as required.

[4] Words referring to the source country's tax system should be explained; they are important as they are ubiquitous in bureaucratic and legal/ administrative language. There is often no equivalent terminology, and students will often have to explain the underlying concept. Compiling a small glossary of such key-words is a good time-saving strategy.

[5] In this case the officer is helpful and kind to the foreign-language speaker. Students should be aware that this is not always the case, and situations like this one may indeed be very stressful for the applicant/client. Indeed, when queues are long at the police station tension and tempers may rise.

Dialogue 12   A charge of manslaughter

SITUATION: A courtroom examination of the defendants. A girl has been found dead on the side of the road. The two foreign defendants have been charged with manslaughter and 'failure to provide assistance'.

IT = Italian speakers (court clerk IT3, lawyer IT1 and magistrate IT2)
E1 and E2 = English speakers (defendants) alternating
Level of difficulty: difficult.
Very specific legal terminology is used and the topic is particularly delicate.

Terms supplied: court-appointed counsel (*avvocato d'ufficio*), examination of the defendant (*interrogatorio dell'imputato*), magistrate for preliminary investigation (*GIP = Giudice per le indagini preliminari*)

IT3:   Buongiorno, Sig. Dabti e Sig. Mugabe. In questo interrogatorio avrete la possibilità di chiarire la Vostra posizione relativamente a quanto è avvenuto la notte del 27 marzo 2010. // Vi presento il Dr. Morelli, Avvocato d'Ufficio. Il GIP, Dr. Valdoni, ascolterà la descrizione di quanto è avvenuto quella notte. //
 La mattina del 28 marzo 2010 alle ore 5.45 il Sig. Rossi, netturbino, dipendente del Comune di Mondolfo, si stava recando in auto al Municipio per iniziare il turno di lavoro alle ore 6.00 quando vide il corpo di una persona sul ciglio della strada, a 1,5 km dal cartello stradale che indica l'inizio del Comune di San Costanzo. // Si è fermato per soccorrere questa persona, ed ha trovato la signorina, che versava in gravi condizioni di salute e in uno stato di confusione mentale. Ha immediatamente chiamato l'ambulanza dal telefono cellulare datogli in dotazione dal Comune di Mondolfo per servizio. // L'ambulanza dell'Ospedale di Mondolfo è arrivata dopo sette minuti e ha portato la giovane donna in ospedale. Sfortunatamente la signorina è deceduta durante il breve tragitto verso l'ospedale. Dopo aver esaminato i documenti rinvenuti nella borsa della giovane donna deceduta, si è ritrovato un biglietto con i vostri nomi e indirizzi.//
 Ora possiamo dare inizio all'interrogatorio. Avvocato, può iniziare quando desidera. *Good morning, Mr Dabti and Mr Mugabe. In this examination You'll be given the opportunity to clarify Your position as to the events that occurred during the night of March 27th 2010.* [1] // *This is Mr Morelli, Your court-appointed counsel.* [2] *Mr Valdoni, Investigating Judge, will listen to the description of the events that took*

place that night. // On the morning of March 28th 2010, at 5.45 am Mr Rossi, street sweeper, employee of the Municipality of Mondolfo, was driving to the Town Hall to start his working shift at 6.00 am when he saw the body of a person lying on the roadside 1.5 km from the road sign marking the beginning of the Municipal district San Costanzo. // He stopped to assist said person and found the young woman, in very poor physical condition and in a state of mental confusion. He immediately called an ambulance from the mobile phone provided to him by the Municipality of Mondolfo. // After seven minutes, an ambulance from the Hospital of Mondolfo reached them and took the young woman to the hospital. Unfortunately the young woman died during the short trip to the hospital. After examining the documents contained in the bag of the deceased young woman, a note with Your names and addresses was found. // Now we can start the examination. Counsel, You may start when You wish.[3]

IT1: Vorrei sapere dagli imputati da quanto tempo sono in Italia e dove risiedono. *I'd like to ask the two defendants how long they have been in Italy and where they live.*

E1: We've been living in Pesaro for four months now. **[4]**

IT1: Sappiamo che siete in possesso di un permesso di soggiorno regolare. *We understand that You have a regular residence permit.*

E1: Yes. A friend of ours, who arrived in Italy in 2007, called us in December, last year; he said that the contractor he was working for was looking for more bricklayers.

IT1: La prego di proseguire con l'interrogatorio degli imputati, Signor Giudice. Io parlerò con i miei clienti al termine dell'interrogatorio. *Please continue examining the defendants, Your Honour. I will talk to my clients at the end of this examination.* **[5]**

IT2: Abitate con questo amico? *Do You live with this friend?*

E2: Yes, we share the same flat in Pesaro.

IT2: Da quanto tempo lavorate? *How long have You been working for?*

E2: We started working in mid-January.

IT2: Dove vi trovavate la sera del 27 marzo? *Where were You on the night of March 27th?*

E1: We went to Xantos, the disco near Marotta.

IT2: Quando e dove avete incontrato la signorina? *When and where did You meet the young woman?*

E1: We met her at the disco that night.

IT2: Perché aveva il vostro indirizzo? Spiegateci che cosa è successo quella sera. *Why did she have Your address? Tell us what happened that night.* **[6]**

E2:   We met her that evening. We'd never met her before. She was a prostitute, working in Marotta. Because we came from the same country, she wanted to make friends with us. She said she felt very lonely here, as two friends of hers had just moved to Rimini – it's busier for them there.

IT2:   Vi ha chiesto di darle un passaggio? A che ora? *Did she ask You to give her a lift? At what time?*

E2:   She didn't feel well and was very tired. I think she'd drunk too much. She felt dizzy and wanted to go back home. She asked us to take her home. It was around 3 o'clock at night.

IT2:   Bene, e dopo che cosa è successo? *I see. And then what happened?*

E1:   We left the disco. I was driving and I had a terrible headache, so I drove slowly, I was afraid something might go wrong. [7]

IT2:   C'è stato un problema? *Was there a problem?*

E1:   Yes, she told us she didn't feel well. She said she was about to vomit. So I stopped the car and let her out. She started vomiting on the roadside.

IT2:   Perché non l'avete portata in Ospedale, al Pronto Soccorso? *Why didn't You drive her to hospital, to the emergency ward?*

E2:   We were really worried, and very confused. He, my friend, had drunk too much. We come from the same village, You know, we're cousins. We came here because our relatives need money. I have two brothers, he has one brother and one sister. They have to go to school. They need the money we send back home. [8]

IT2:   Dunque, torniamo a quello che è successo. Allora, la signorina non si sentiva bene. E voi la lasciate lì, sul ciglio della strada, di notte, in una strada dove passano poche auto, o non passa proprio nessuno? Vuol dire abbandonarla al suo destino, in pratica vuol dire condannarla a morte. *Well, let's go back to what happened. Let me get this right, the young lady didn't feel well. It was very early in the morning and You leave her there, at night, on the side of a road where there is no traffic, or almost no traffic at all. Abandoning her, practically sentencing her to death.*

E1:   No, no; we didn't want to hurt her, we were just confused. What do you mean? We're not killers. We didn't know her, we had nothing against her. We had just arrived in Italy, we didn't know anybody, we just wanted to enjoy ourselves a bit, it was Saturday evening ... Then she didn't feel well, we didn't know what to do, we were so worried the police might think we were in business with her ... [9]

IT2: Ascoltate. Adesso sappiamo che cosa è successo, e l'indagine accerterà i fatti. Resterete in stato di detenzione al Carcere di Villa Fastiggi fino a che la Corte non avrà preso una decisione. L'avvocato d'ufficio, il Dr Morelli, vi seguirà. Parlate con lui per decidere la vostra difesa. La seduta è tolta. *Listen. Now we know what happened and the facts will be ascertained through further investigations. You will be held in custody at the Villa Fastiggi prison until the court has made its decision. Dr Morelli, Your counsel, will assist You. Talk to him now to discuss Your defence. The hearing is adjourned.*

Commentary

[1] This very formal description of events allows trainers to immediately tackle the difficulties found in this high register and formal language. This passage is so long – respecting the Italian legal procedure – that it needs to be broken into several chunks, according to the students' level and individual abilities. It is so long and dense with information that it also works well as an opportunity to practise interrupting strategies, i.e. that the students themselves learn to stop the speaker when they feel they can no longer retain the information in their short-term memory. Because of the length, the presence of names, times, dates, distances, places etc., it is also well-suited for notetaking practice.

[2] A good opportunity to discuss legal terminology and legal professions.

[3] This is a perfect opportunity to discuss police statements and reports, police procedure and the chronology of a case from arrest to sentencing or acquittal, as well as the various actors involved in legal procedures.

[4] This dialogue allows trainers to discuss several important issues: English being used as a lingua franca, and difficulties in fully understanding the defendants' strong accent (that may be simulated by the teacher or used for listening practice with a recording, not of the trial, but of non-standard English spoken in a different situation, maybe a TV or radio programme or interview found on the internet).

[5] The dialogue must of course be adapted in those passages where the primary interlocutors are addressing each other. This is an excellent opportunity to discuss – both in the light of the code of ethics as well as the current practice in the L1 culture – exactly how much the interpreter is expected to translate of the conversation, and if there are passages that do not need translating (e.g. conversations between the primary interlocutors or other passages that the interpreter is not expected to translate – again in the light of the code of ethics and in the light of expectations of the L1 culture).

[6] These question-answer sessions are excellent practice for question formation and for using the imperative (in bald or mitigated forms: 'tell us', 'please tell us', 'I'd like to know more about', etc.). They are also good for discussing question formats – see chapter 5.

[7] It is very important to stress that the interpreter has to be accurate, and translate everything, using the same register and lexicon.

[8] The issue of impartiality can be discussed here; a situation like this may trigger processes of bonding and empathy that need to be handled carefully. These situations can actually be more difficult to deal with and less straight-forward – also for the legal actors – than they sound 'on paper', and it is good for students to have aired these issues with trainers and peers before they find themselves involved in these situations and having to make painful ethical decisions. This passage isn't particularly dramatic, but there are times when witnesses, defendants, applicants for political asylum, etc. break down and cry when relating their life stories (which may or may not be truthful). Police officers, lawyers and magistrates are trained to deal with these situations, and interpreters too should be prepared to tackle them.

[9] The defendant is very anxious and he is speaking fast, with a strong accent, and stressing that he is not guilty. This is a perfect opportunity to introduce the obstacles, suggested in chapter 5, in the form of emotional stress.

The rapid succession of questions, like the patient's questions in Dialogue 8 (see comment 12 there), is of course an indication of the speaker's frame of mind and would be spoken quickly in an uninterrupted flow. Trainers and trainees will have to decide how to divide this chunk and manage the turns. Discourse markers will sometimes be left untranslated because there is no obvious correspondence, but students should be encouraged to look for compensatory strategies, since discourse markers can be used as argumenta-tive and confrontational devices (Hale 1999), expressing challenges and disa-greement during cross-examination and controlling the progression of talk.

Dialogue 13   A charge of assault and bodily harm

SITUATION: An American citizen is arrested on a charge of assault and bodily harm – a summary trial.

IT = Italian (magistrate)
E = English (defendant)
Level of difficulty: very difficult.
This dialogue contains complex legal terminology as well as naturally occurring errors in the speaker's re-telling of the events, portraying a realistic mix of legal terminology and spontaneous speech marked by fear and confusion.
Terms supplied: bodily harm (*lesione personale*), injured party (*parte lesa*), summary trial (*processo per direttissima*). There is often no clear equivalence in the two legal systems, and words may have to be explained in the target language. Many more terms could be provided here, adapting the exercise to the level of the students' abilities and the trainer's pedagogical aims.

IT: Procediamo a convalida dell'arresto. Chiedo le generalità dell'imputato con ammonizione che in caso di rifiuto di fornirle, o fornirle false, Lei sarà punito ai sensi degli artt. 495 e 496 c.p. con la reclusione anche fino a tre anni. E' chiaro? *The defendant is under arrest* [1]. *You must provide your personal details and You are warned that if You refuse to give them, or supply false details, You will be convicted, according to art. 495 and 496 of the Criminal Code, to up to three years in prison. Is that clear?*

E: Yes.

IT: Ci dia le sue generalità: nome, cognome, luogo e data di nascita. *Please state Your personal details: name, surname, place and date of birth.*

E: David Newman, I was born in Seattle, USA, on January 16th, 1981.

IT: Ci mostri un documento d'identità. *Show us Your identity card, or passport.*[2]

E: I have my driving licence. I came to Italy a week ago, I'm just staying for a couple of weeks as a tourist.

IT: Adesso ci deve fornire i seguenti dati: residenza anagrafica – stato civile – professione – precedenti condanne o carichi pendenti. *Now please give us the following information: Your permanent address, if You are married, Your profession and if You have a criminal record.*

E: I don't have an address here in Italy, I'm just here as a tourist. I'm travelling in a camper, I live with my girlfriend, I'm a freelance journalist, I have no criminal record.

IT: Signor Newman, a questo punto la invito ai sensi dell'art. 161 comma 1 c.p.p. ad eleggere domicilio per le notificazioni del presente procedimento penale. In mancanza, le notificazioni verranno eseguite mediante consegna di copia dell'atto al Suo difensore d'ufficio Avv. Penna del Foro di Forlì. *Mr Newman, I invite You, pursuant to art. 161 paragraph 1 of the Criminal Code* [3], *to choose a domicile so that You can be notified regarding these criminal proceedings. Should this not be forthcoming, You will be notified by delivery of the deed to Your court-appointed counsel, Mr Penna, registered lawyer in Forlì.*[4]

E: I'm staying with a friend now, Sara Rossi, in Via Regnoli, 25, Forlì.

IT: La informo del motivo per cui è stato arrestato: ipotesi di reato di 'Minaccia aggravata (art 612 c.p.) e Lesione personale (art. 582 c.p.) e per tutti quei reati che verranno ravvisati nei fatti esposti'. E' chiaro? *The reason You have been arrested is: 'Aggravated assault (art. 612 of the Criminal Code) and bodily harm (art. 582 of the Criminal Code) and for all the crimes that will be ascertained in the facts described'. Is this clear to You?*

E: Yes.

IT: Procediamo all'interrogatorio dell'arrestato ai sensi dell'art. 391 comma 3 c.p.p. La avverto che, se intende rispondere alle domande che le verranno fatte le sue dichiarazioni potranno essere sempre utilizzate nei suoi confronti. // Lei ha la facoltà di non rispondere ad alcuna domanda, ma il procedimento seguirà comunque il suo corso... se renderà dichiarazioni su fatti che riguardino la penale responsabilità di terze persone assumerà, riguardo a tali fatti, l'ufficio di testimone. Intende rispondere? *Now the arrested person will be questioned pursuant to art. 391 paragraph 3 of the Code of Criminal Procedure. I warn You that if You intend to answer the questions put to You, what You say can always be used against You. // You are entitled not to answer the questions, but the proceedings will continue regardless ... if You make any statement involving the criminal liability of third parties You will, with regard to said facts, act as witness. Do You intend to answer my questions?*

E: Yes, I do.

IT: Racconti le circostanze che hanno portato alla colluttazione e al ferimento della parte lesa (Francesco Rossi). *Tell us about the events leading to the fight and the wounding of the injured party (Francesco Rossi).*

E: I'm innocent. Last night, yesterday, I went down to the pub and I was hanging out with some friends. We had something to drink together and then we decided to go to the beach. We split up into two groups, we took a car each, and I joined one group in one of the two cars.

IT: Come si chiama il pub, come si chiama la persona che guidava l'auto e che tipo di auto era? *What was the name of the pub, and what was the name of the person driving the car and which type of car was it?*

E: The pub was called The Shamrock, I think, something like that. The driver's name was Dario, I have no idea what his surname was. The car looked a lot like a Golf to me. Yeah, a Golf.

IT: E poi cosa è successo? *And then what happened?*

E: On the motorway the two drivers began racing, the other car was challenging us and once we were out, soon after the Rimini toll station, the other car cut us off and almost caused an accident. // The driver of the car I was in started beeping the horn. He was really angry – we'd almost tipped over! Then he overtook the other car and blocked it. Both drivers came out of the cars and started yelling at each other. // I couldn't understand what they were saying, but I told Dario – he speaks English – not to worry, it's not a big deal, don't get out of the car, let him go. // I was getting kind of frightened. But he went up to the other guy and they started shoving each other around. So I tried to break up the fight.

IT: Spieghi la dinamica più dettagliatamente. *Explain to us what happened in more detail.*

E: I went to break up the fight and I was hit. The other driver started swearing, or that's the way it sounded to me, and I told Dario to leave him alone.

IT: E come spiega che invece la parte lesa ha riconosciuto lei come il suo aggressore? *How do You explain the fact that the injured party has recognized You as his assailant?*

E: I tried to divide them, and at the same time to protect Dario, the guy I was in the car with.

IT: Ho qui un verbale di sequestro di un coltello a serramanico a scatto con lama da 10 cm, manico di osso marrone. Questo coltello è suo? Lo riconosce? *I have here the seizure report: a jack-knife with a 10-cm blade, and a brown bone handle.* [5] *Is this knife Yours? Do You recognize it?*

E: Yes, it's mine, but I didn't use it!

IT: Ha una licenza per portare quest'arma? *Do You have a licence for this weapon?*

E: No, but in the USA You don't need a licence for this! [6]

IT: Qui non siamo in America. Comunque, la parte lesa, Francesco Rossi, ha dichiarato, e leggo dal verbale di denuncia rilasciata ai carabinieri: 'Riconosco David Newman come il mio aggressore. Durante la colluttazione ha tirato fuori dalla giacca il suo coltello, brandendolo e poi mi ha colpito al volto'. Cosa ha da dire al riguardo? *We are not in the United States here. In any case, the injured party, Francesco Rossi, stated – and I'm reading the report of the crime [7] filed with the police: 'I recognize David Newman as my assailant. During the fight he took the knife out of his jacket, brandished it and then hit me on the face.' What can You tell us in connection with this?*

E: He must have been confused. I have a knife with me because I'm travelling alone and I want to be able to protect myself in case of trouble. But the knife fell out of my jacket while I was trying to separate the two guys. It was Dario who took it and injured Mr Rossi.

IT: Ora, mi racconti le circostanze del suo arresto. *Now, tell me what happened when You were arrested.*

E: The Carabinieri [8] arrived on the scene just when Mr Rossi's face got hurt... they asked me if the knife was mine, and I said yes, but I wasn't holding the knife then. I tried to explain to them what had happened, but they handcuffed me and immediately brought me to the Carabinieri station without letting me explain what had happened. Then Dario must have blamed me for the assault and must have taken advantage of the fact that I don't speak your language very well!

IT: Ok, ha altro da aggiungere in sua difesa? *Ok, do You have anything to add in Your defence?*

E: Yeah, I'm a decent guy, I only have a knife on me to protect myself. Dario was a bit drunk when we left the pub. His friends can confirm that they didn't want him to drive the car.

IT: Quali amici? Ci fornisca i loro nomi e ci dia una loro descrizione fisica. *Which friends? Give us their names and describe what they look like.*

E: I only remember the name of one guy: Roberto. He's slim, 5 feet 3 inches [9], black-haired.

IT: Ok. Basta così. Si chiuda il verbale. *Ok. That will do. The report is closed.*

IT: Visti gli art. 391 e 549 c.p.p. si convalida l'arresto del Sig.Newman in quanto avvenuto nella flagranza del reato di Lesione personale (art. 582 c.p.) come risulta dal verbale di arresto in data 14.09.2010 e del reato di porto abusivo di armi di cui all'art. 699 del codice penale in quanto l'indagato, per sua stessa ammissione, era in possesso di un'arma senza licenza. // Ritenuto che nella fattispecie sussistano a carico del Newman gravi indizi di responsabilità in ordine ai reati a lui contestati, ritenuto che sussistano nella specie anche esigenze cautelari che impongono l'adozione di una misura restrittiva, in particolare che ricorre il pericolo di fuga, questa corte dispone nei suoi confronti la misura coercitiva personale degli arresti domiciliari da eseguirsi presso l'abitazione sita in Forlì, Via Regnoli 25 e ne ordina l'immediata liberazione. Dispone procedersi con rito direttissimo. *Pursuant to articles 391 and 549 of the Code of Criminal Procedure we confirm the arrest of Mr Newman* **[10]**, *who was caught in the act of inflicting bodily harm (art.582 of the criminal code), as stated in the arrest report dated September 14th, 2010, and illegal possession of a weapon, according to art. 699 of the criminal code, as the defendant by his own admission had a weapon without the necessary weapons licence. // Considering that there is substantial evidence against Mr Newman as to the crimes he has been charged with* **[11]**, *and that custody is required as there is a risk that the defendant may flee, this court orders the defendant to be subject to house arrest at his place of residence, in Forlì, Via Regnoli 25 and orders the same to be immediately released. The court will commence the summary trial.*[12]

Commentary

[1] Legal terminology, and the relevant expressions, can be addressed here: '*arrest*', '*house arrest*', '*confirmation of arrest*', etc.

[2] The tone of the questioning is far more aggressive in this dialogue than in the first dialogue. Students should note how questioning strategies and features such as modality, hedging, the use of bald imperatives, change along with the change in tone.

[3] Basic acronyms need to be explained. In this case the Italian c.c.p. stands for Code of Criminal Procedure, and c.p. for Criminal Code.

[4] Students should be reminded of the various actors and procedures in the legal system and issues such as the defendant's rights, namely having a court-appointed counsel if the suspect doesn't have his own lawyer.

[5] Describing an object like a knife may pose some difficulties. Students should be encouraged to be as accurate as possible when describing evidence.

[6] The different legal provisions regulating the possession of weapons can be discussed here, as there are big differences between the US and other countries in this respect.

[7] Various types of crimes can be discussed here.

[8] *Carabinieri*, the 'army police', is untranslatable and can be left as it is in this context. Several legal terms have no exact equivalent in the other language and have to be kept in the original language. Students are encouraged to draw up a list of these terms.

[9] Converting units of measurement is always a difficult task. Students can prepare a small table of measurements listing men's and women's average heights and weight to use when needed.

[10] This long reading of Mr Newman's arrest confirmation requires a whispered translation or a consecutive interpreting mode if the defendant is not seated next to the interpreter. If practising chuchotage, students can try to translate one sentence at a time. Trainers should insist on the requirement to ask for interruptions or repetitions whenever the need arises.

Although this translation is quite close, it has not maintained the formality and very high register of the source text. In these cases, the ideal procedure is to use parallel texts: to find corresponding documents in the L2 system and try to aim for a style, register and lexis that is as close as possible to that used in the L2 system without, however, sacrificing any of the information contained in the source text. Especially in a context such as this, where so much is at stake for the defendant and where the judge needs as precise information as possible, accuracy takes precedence over style, register, idiomatic language and 'naturalness'.

[11] Again, another opportunity to discuss terms like *'charge'*, *'indictment'*, etc. along with differences in arrest procedures in the two legal systems.

[12] This is another case of differing procedures that needs to be explained to the students, stressing the need to explain different rules that very often cannot be translated literally, but rather have to be explained in the target language.

Dialogue 14   Interpreting for a burglary case

SITUATION: An Italian woman is arrested by the police in England and accused of burglary. The interpreter is present at the first meeting between the woman and her English lawyer.

IT = Italian speaker (the defendant)
E = English speaker (English lawyer)
Level of difficulty: quite difficult.
Very specific legal terminology is used and the students should have a general knowledge of both legal systems, as several differences emerge.
Terms supplied: burglary (*violazione di domicilio con furto*); court-appointed counsel (*avvocato d'ufficio*); to plead guilty (*dichiararsi colpevole*); breaking and entering (*furto con scasso*), premeditated (*premeditato*).

E: According to the charge sheet here, you entered the premises of Brown's, 63 Brompton Road, London, on July 1, 2010, and took away a handbag containing a purse with £368 in cash, three credit cards, a chequebook, a Switch card, and a mobile phone. Did the police caution you when they arrested you?

IT: Che cosa significa, sta parlando dei miei diritti? Mi hanno detto che avevo il diritto di restare in silenzio. *What do You mean, are You talking about my rights? I was told I was entitled to remain silent.*

E: Before asking any questions, the police must tell you that you have the right to remain silent, and that anything that you say may be used against you in court, and that you have the right to consult with a lawyer who can be present during questioning. [1]

IT: Sì, me l'hanno detto e mi hanno detto che avevo diritto ad un avvocato d'ufficio e hanno chiamato lei. Non mi hanno ancora interrogato. *Yes, they told me, they told me that I was entitled to a court-appointed lawyer and they called You. I've not been questioned yet.*

E: When the police interview you, before you answer I'll advise you whether you should be answering or whether you should remain silent.

IT: Capito. Allora adesso cosa succederà? Mi interrogheranno? E poi, mi porteranno in prigione? Sono così preoccupata ... *Ok. What's going to happen now? Will I be questioned? And will they put me in prison? I'm getting a bit worried...*[2]

E: Probably not. We can apply for bail, so that you can be released. Since it's not a serious offence, there shouldn't be any problems.

IT: Che significa, rilasciare su cauzione? Mi rimettono in libertà? E quando ci sarà il processo? Devo restare molto a lungo qui in

Inghilterra? Vorrei tornare a casa in Italia prima possibile. *What does that mean, applying for bail? Will I be released? And when will the trial start? Do I have to stay here in England for long? I'd like to go back home to Italy, as soon as possible.* **[3]**

E: To release someone on bail means you're allowed to be at liberty, rather than being held in custody, while awaiting the next stage in the criminal process, i.e. when you have to appear before the Magistrate's Court, which is where your case will be heard. // That will probably be in about three days' time. The police will inform us after they interview you and hopefully they will have granted you bail.

IT: In Italia i processi durano anche anni, qui come funziona? Non voglio restare qui bloccata per tanto tempo, voglio tornare a casa mia dopo l'estate. *In Italy trials can take years, what happens here? I don't want to be stuck here for long; I'd like to go back home at the end of summer.*

E: Since it's not a serious offence, unless you choose to be tried by a jury at the Crown Court **[4]**, which I don't advise as you were caught in the act – you'd just get a tougher sentence for wasting the Court's time, the case will be dealt with in the Magistrates' Court. Then, since, as I said, you were caught red-handed, it's in your best interests to plead guilty.

IT: E se mi dichiaro colpevole, che genere di pena mi posso aspettare? Dovrò andare in carcere? *And if I plead guilty, what kind of sentence can I expect? Will I be imprisoned?*

E: That depends. Did you break any windows or force any windows or a lock to get into the house?

IT: No, stavo distribuendo della pubblicità, è un lavoretto part-time che ho trovato, sono una studentessa e sono molto a corto di soldi. // Ho visto che la finestra era aperta, c'era una borsa sul tavolo. Non c'era nessuno e sono entrata facilmente. Ma non ci sono andata con l'intenzione di rubare. La tentazione è stata troppo forte. *No, I was distributing advertisements – it's a part-time job I have – I'm a student and I'm always broke. // I saw the window was open, and there was a bag on the table. There was no one inside and it was easy to enter the house. But I didn't go there intending to steal anything. It was just that the temptation was overwhelming.*

E: Good, in that case there's no breaking and entering, and it wasn't premeditated, which would have made the offence much more serious. Probably in this case the magistrates will not impose a custodial sentence, a fine will be sufficient.

IT: Mi rimanderanno a casa in Italia? Ero venuta per imparare la lingua e non vorrei andarmene via ancora per un po. // Ma la multa che dovrò pagare quanto sarà? Come le ho detto, non ho molti soldi. *Will I be sent back home, to Italy? I came here to study English and I'd like to stay a little longer. // How much will the fine be? Like I said, I don't have much money.*
[The lawyer raises his hands in despair and takes his leave....]

Commentary

[1] Miranda rights can be discussed before starting the dialogue. One might also remind the students that cautions vary from one jurisdiction to another. The interpreter's role can also be discussed here. If the defendant does not understand the lawyer, the interpreter should immediately inform the latter and should not try to explain legal concepts to the defendant.

[2] If the defendants or witnesses are clearly frightened, it is essential to address the actors correctly and appropriately, as fear may lead to aggressive behaviour. If the defendant is worried he may start asking questions, and might not even allow the interpreter to fully translate the lawyer's comments before asking a new question. The interpreter should be firm and assertive here with the turn-taking procedure.

[3] Here too we have another string of questions and an anxious interlocutor.

[4] This is a legal term that cannot be directly translated into Italian, as the legal systems existing in the two countries are different. The interpreter should leave the English term and be familiar with the type of offences that are tried there. The same applies to Magistrate's Court.

Dialogue 15   Interpreting for a breaking and entering case

SITUATION: Mr Decker testifies in a breaking and entering case. The interpreter is asked to assist the witness and the defence lawyer.

IT = Italian speaker (defence lawyer)
E = English speaker (Mr Decker, witness)
Level of difficulty: medium.
Specific legal terminology is used, and the description of events is long and rich in details, including the physical description of the burglar.
Terms supplied: breaking and entering (*furto con scasso*), burglar (*ladro*), expert witness (*perito*), eye-witness (*testimone oculare*), re-enactment (*simulazione, ricostruzione*).

IT: E' esatto dire che Lei ha conosciuto il mio assistito, il signor Martini il 2 luglio 2010, quando è arrivato alla villetta che aveva preso in affitto per le vacanze estive con sua moglie, tale Peggy Sue Decker? *Is it correct to say that You met my client, Mr Martini, on July 2nd, 2010, when You arrived at the villa You had rented for the summer vacation with Your wife, Mrs Peggy Sue Decker?*

E: Yes, that is correct.

IT: Ed è esatto dire che il furto da Lei subito è avvenuto il 31 luglio dello stesso anno? *And is it correct to say that You were robbed on July 31st of the same year?*

E: Yes, that is correct.

IT: Quindi immagino che in tutto questo tempo ci siano state occasioni, in cui lei o sua moglie abbiate frequentato il suo vicino di casa, il mio assistito signor Martini, in cui lo ha invitato a casa sua? A bere qualcosa, non so un aperitivo? *Therefore, I suppose that during this period when You or Your wife used to meet Your neighbour, the defendant, Mr Martini, you had the opportunity to invite him to Your house? For something to drink, for a drink before dinner, maybe?*

E: No, never.

T: Come mai? *Why?*

E: I didn't like him. Instinctively I just didn't like him. He's kind of creepy! Every time my wife and I came back home or came out, there he was, on the veranda, staring at us, or spying on us! I didn't trust him at all, and he got on my nerves!

I: Signor Decker, a sommarie informazioni il 31 luglio 2010, lei ha dichiarato, e poi confermato qui durante l'esame del PM, di essersi svegliato con un tonfo che, in base alle perizie della scientifica, [1] con ogni probabilità era la ringhiera della finestra che veniva forzata. // Poi ha sentito un altro rumore, quello del vetro della finestra della cucina che veniva rotto. E' esatto? *Mr Decker, in the statement made on July 31st 2010, You declared, and then confirmed here during the Prosecutor's examination, that You were woken up by a thud that – based on the forensic police report – was most probably due to the window bars being broken open. // You then heard another noise, that of the kitchen window being broken. Is this correct?*

E: Yes.

I: Poi cosa è successo? *What happened next?*

E: At that point I got up, grabbed one of my golf clubs as a weapon; they were next to the wardrobe in my bedroom, and I went to where I had heard the noise. I looked out of the door of my room and I saw the thief putting something in a small bag.

I: Ora quale crede che sia il lasso di tempo intercorso dal momento in cui ha sentito il vetro rompersi e il momento in cui ha visto il ladro? *How much time do You think elapsed from when You heard the window being broken and when You saw the burglar?*

E: I don't know, 4 or 5 seconds; I was already awake at that point, and I immediately reacted – you know, I was already on the alert because of the previous sound.

I: Allora 4 o 5 secondi. In questo lasso di tempo il ladro quindi ha scavalcato la finestra della cucina, ha attraversato il corridoio, è entrato nel salotto dove ha aperto il cassetto, ha estratto i soldi e li ha infilati in una borsa! Giusto? [2]. *In the time elapsed the burglar climbed through the kitchen window, crossed the corridor, entered the living room where he opened the drawer took out the money and put it into his bag! Is that correct?*

E: What do you mean?

I: Signor Decker, noi abbiamo provato a fare questo stesso percorso nella villetta da lei presa in affitto sapendo esattamente dove dirigerci, e quindi conoscendo la dislocazione delle stanze della sua villetta e ci abbiamo messo sette secondi. Quindi è evidente che questo ladro sapesse già cosa cercare e dove cercarlo. // Ora come crede sia possibile che il mio assistito che, come lei ha testé affermato, non è mai entrato in casa vostra, che, per sua stessa ammissione, non era un vostro amico e quindi non poteva sapere che avevate nascosto i soldi nel cassetto, come è possibile secondo lei che il mio assistito abbia impiegato 4 o 5 secondi a fare tutto questo? [3] *Mr Decker, we tried to cover the same distance in the villa that You rented, knowing exactly where to go and where the rooms were located, and it took us 7 seconds. It is therefore clear that the burglar already knew what he was looking for and where to find it. // Now, how could my client, who – as You have just stated – had never been in Your house, and who, as You've just told us, was not a friend of Yours – know that You had hidden the money in the drawer? How is it possible for my client to do all this in just 4 or 5 seconds?*

E: I don't know, maybe I'm wrong ... maybe it took me longer to come out of my bedroom...

I: Lei ha appena detto, e se vuole le faccio rileggere il verbale: 'Ero già sveglio, ho reagito immediatamente ero già allertato dal rumore precedente.' Vuole risentire le sue parole? *You've just said – and if You like I can read from Your statement: 'I was already awake at that point, and I immediately reacted – You know, I was already on the alert because of the previous sound.' Would You like me to repeat Your words?*

E: No, that's not necessary. I don't know how he knew where the money was. He was probably spying on us through the window when we were at home. I just *know* that the person who was stealing in my house was Mr Martini!

I: A questo punto vorrei avanzare una contestazione. Signor Decker, nella sua denuncia rilasciata la sera stessa del delitto, 31 luglio scorso, lei ha descritto l'uomo che si era introdotto nella sua casa, in questo modo: 'Ho visto l'uomo da dietro, era di altezza media circa 1.75, snello però si muoveva in modo goffo, barcollava un po', indossava un cappello con visiera come quelli da baseball'. Non un uomo robusto, alto 1.85 e pelato. *I dispute Your statement then. Mr Decker, in the statement that You made on the same evening the burglary took place, that is July 31st last year, You described the man who had sneaked into Your house as follows: 'I saw the man from behind, medium height, 5 feet 7 approximately, he was a slim man but he walked clumsily, he swayed a little, and he wore a cap like a baseball cap.' You did not describe a well-built, 6 foot, bald man.*

E: But that evening I was upset, I was in shock, you know, due to the burglary and all that... However I recognized Mr Martini later, during the line up at the *Carabinieri* station. I was not 100 per cent sure at first, but then I realized it was him.

I: Non ho altre domande. Signor Giudice, a questo punto vorrei formulare una richiesta a questa Corte. Considerate le numerose contraddizioni in cui è caduto il teste durante l'esame, considerato che il Decker risulta essere l'unico testimone oculare del delitto, nonché principale teste a carico del mio assistito, la difesa richiede che venga ammesso come teste per la difesa il consulente tecnico che ha condotto la simulazione del furto commesso nella villetta, il Dottor Romolo Sandri. *I have no further questions. Your Honour, at this point I would like to put a request to the Court. Considering that the witness has contradicted himself on several points during the examination, and considering that Mr Decker is the only eye-witness to the crime, as well as the main witness against my client, the defence asks the Court to admit Romolo Sandri – the expert witness who directed the re-enactment of the burglary committed at the villa – as a witness for the defence.*

Commentary

[1] This dialogue contains dense legal terminology. Students often find it useful (and the fact that it is entertaining increases motivation...) to watch TV detective and courtroom series such as *CSI* and *My Cousin Vinny*, a courtroom drama, to improve their knowledge of legal jargon and, not least, to improve their comprehension skills.

[2] This very precise description of events allows trainers to stress how important it is to be accurate and report every single detail of an account.

[3] When translating a sequence of events like this, remind the students to take note of the core words in the chronology.

Notes

Chapter 1

1. Although there are various other alternatives, we use the terms 'service provider' and 'client' to describe the interlocutors. By 'service provider' we mean the representative of the host institution (doctor, nurse, police officer, judge, etc.) and by 'client' we mean the person seeking the service provided by that particular institution, who does not speak – or speak sufficiently well – the language of the host country (e.g. a patient, tourist, defendant, refugee status applicant, and so forth). By 'host country' we mean the new country in which the activity is taking place and in which the person (migrant, tourist, businessman) is currently staying or living.
2. We use the singular form 'community' here for simplicity, but the pluralistic nature of migrant communities should be presumed.
3. Although this discussion is far beyond the scope of this book, the terminology of equivalence is important and should be given due weight in the classroom.
4. Gideon Toury's ground-breaking translation model, first proposed in the early 1980s, explains this process very well and has impacted deeply on the development of the discipline. Indeed, Translation Studies has been largely devoted to addressing questions of equivalence both at a descriptive and at a prescriptive level. Anthony Pym's (2010) *Exploring Translation Theories*, Jeremy Munday's *Introducing Translating Studies; Theories and Applications* (2008), and Franz Pöchhacker's (2004) *Introducing Interpreting Studies* offer a very useful panorama of Translation Studies to the present day.

Chapter 2

1. To speech must be added culturally defined non-verbal communication such as body language and gesture. Two very simple, humorous and informative books that are popular with the students are *Gestures: The Do's and Taboos of Body Language Around the World* and *Do's and Taboos Around the World* by Roger Axtell (1991, 1993). There are innumerable websites that offer information on cross-cultural differences in body language (e.g. http://www.businessballs.com/body-language.htm) and how to avoid misunderstandings due to these differences.
2. This was the claim made by one of the earliest translation scholars, Eugene Nida (1964), who argued that the aim of translation is to create the same effect in the target reader through the translation that the original text had on the original reader.
3. The hierarchy can be so daunting for the patients/suspects/witnesses that the interpreter might feel the need to mitigate its effect through tone of voice, body language or politeness formulae, simply in order to put the

foreign-language user in a position to speak rather than not speak out of fear. The ideal solution is of course to signal this communication-hampering situation to the service provider, but for a number of practical reasons it may not always be possible.

4. This statement is problematic in that it is impossible for any one person to be fully cognizant of another's person's intent, and it is equally impossible to be sure that one has fully comprehended, let alone is able to successfully transfer, that assumed 'speaker's intent'. This translation process will depend on the interpreter's ability to interpret the interlocutor's utterances according to what is generally recognized as the dominant cultural norms to the degree that this is at all possible. Clearly, it will also depend on the interpreter's own interpretation of the interlocutor's intention and on her interpretation of the cultural and language norms. It is important to remember therefore that these interpretative processes are subjective and variable. This dilemma is at the heart of translation theory and has been debated hotly from its inception to the present day. The trend in Translation Studies has indeed been to move away from a positivist understanding of the text whereby it is possible to decipher and transfer an assumed monolithic meaning, to a far more realistic approach that recognizes the enormous challenges inherent in this endeavour. Nevertheless, the very process and nature of written and oral translation is based on the act of interpreting a text and transferring it (form, content, assumed meaning) into another language form. Interpreting and Translation Studies must therefore accept that the aim and objective of the practice is unachievable in absolute terms but, significantly, that must not prevent it from being pursued.

5. We must not forget that interpreters have their own personal history and trajectory through which their own identity is construed. They can never entirely overlap culturally with their interlocutors (culture after all is composed of a host of other factors as well as ethnicity) any more than any two individuals in the world completely 'overlap'.

6. There is a host of intercultural communication books available through the internet for teachers, business managers, doctors, etc., many of them catering for specific sectors, and trainers can choose books that meet their own specific needs and the students' levels. For example, the National Centre for Cultural Competence of Georgetown University Centre for Child and Human Development promotes a model for achieving cultural and linguistic competence based on the work of Cross *et al.* (1989) http://nccc.georgetown.edu/foundations/frameworks.html. We have used Holliday, Hyde and Kullman (2004) *Intercultural Communication: An Advanced Resource Book* in the classroom, as it is a useful combination of texts on cultural identity and practical exercises that has suited the level of our students. As for more academic books, again there are many to choose between and here we only mention two classic studies: Scollon and Wong Scollon's (1995) *Intercultural Communication: A Discourse Approach* and Spencer-Oatey and Franklin's (2009) *Intercultural Interaction: A Multidisciplinary Approach to Intercultural Communication.* Many swear to M. J. Bennett's work (see Bennett 1998). We have found Bowe and Martin's (2007) volume very useful in our classwork as it presents established linguistic and conversational discourse theories from the angle of cross-cultural communication, providing a wealth of valuable examples (with

a strong emphasis on politeness theories). The connection culture–language and the problematizing in a cross-cultural perspective of some of the most established theories (Grice, Brown and Levinson, for example) render it a particularly valuable tool for interpreter trainers.

Chapter 3

1. Fox (2004: 152–3) provides an unexpected, thought-provoking and humorous take on patient–doctor communication in England. She claims convincingly that the 'open, communicative' approach encouraged by modern medical associations with direct eye-contact, makes English patients feel uncomfortable rather than putting them at their ease, which is clearly the aim of this agenda, and may hinder rather than encourage the exchange of information.

Chapter 5

1. These situations may seem a bit forced, but humour tends to make group work easier and using situations like these allows us and the students to introduce pragmatic interpersonal features such as culturally defined institutional hierarchies which govern politeness strategies between doctor and patient.
2. We would like to thank our colleague Michela Giorgio Marrano for the ideas expressed in this paragraph.
3. '[A]n item in a court interpreter's target text which has no precedent on the surface of the original utterance. An addition is considered to have no precedent when its presence may not be explained by reference to the interpreting process, or by identifying it as constituting an interpreter error' (2002: 239).
4. Suggested topics for the interpreting module might include:
   - Ethics in community and public service interpreting.
   - Interpreter roles: the court interpreter vs. the health interpreter.
   - A literature review of the field (choose one area).
   - The role of the interpreter can be identified with that of a 'translation machine'.
   - Should codes of ethics be considered as a guideline only?
   - 'The community interpreter should be invisible.' Discuss.
   - 'Business interpreting raises special questions for the impartiality of the interpreter.' Discuss.
   - Comment on a case study of an interpreted dialogue.

Chapter 6

1. We would like to thank Anthony Mitzel for his contributions to Dialogues 6 and 11, and Isabella Preziosi and Chris Garwood for allowing us to adapt and use their scripts for Dialogues 13, 14 and 15.

Bibliography

Angelelli, C. (2004a) *Medical Interpreting and Cross-Cultural Communication.* (New York and Cambridge: Cambridge University Press).

Angelelli, C. (2004b) *Revisiting the Interpreter's Role: A Study of Conference, Court, and Medical Interpreters in Canada, Mexico, and the United States.* (Amsterdam/Philadelphia: John Benjamins Publishing Company).

Angelelli, C. (2007) 'Assessing medical interpreters: the language and interpreting testing project'. *The Translator,* 13 (1): 63–82.

Angelelli, C. and Jacobson, H. (eds.) (2009) *Testing and Assessment in Translation and Interpreting Studies. A Call for Dialogue between Research and Practice.* (Amsterdam/Philadelphia: John Benjamins Publishing Company).

Axtell, R. (1991) *Gestures: The Do's and Taboos of Body Language Around the World.* (Book Review) (New York: John Wiley and Sons).

Axtell, R. (1993) *Do's and Taboos Around the World.* 3rd edition. (New York: John Wiley and Sons).

Baistow, K. (1999) 'Dealing with other people's tragedies: the psychological and emotional impact of community interpreting'. Paper presented at The 1st Babalea Conference on Community Interpreting in Vienna, November 1999, printed privately 2000.

Barsky, R. (1996) 'The interpreter as intercultural agent in Convention refugee hearings'. *The Translator,* 2 (1): 45–63.

Bennett, M. J. (1998) *Basic Concepts of Intercultural Communication. Selected Readings.* (Yarmouth: Intercultural Press).

Berk-Seligson, S. (1990) *The Bilingual Courtroom: Court Interpreters in the Judicial Process.* (Chicago: University of Chicago Press).

Betancourt, J. R. and Cervantes, M. C. (2009) 'Cross-cultural medical education in the United States: key principles and experiences'. *Kaohsiung Journal Med. Sci.,* 25(9): 471–8.

Bhatia, V. K. (1993) *Analysing Genre: Language Use in Professional Settings.* (London: Longman).

Bischoff, A. and Loutan, L. (1999), *Due Lingue, Un Colloquio. Guida al colloquio medico bilingue ad uso di addetti alle cure e di interpreti.* (Ticino: Dipartimento delle Opere Sociali).

Bischoff, A. *et al.* (2003) 'Improving communication between physicians and patients who speak a foreign language'. *The British Journal of Medical Practice,* 53: 541–6.

Bowe, H. and Martin, K. (2007) *Communication Across Cultures: Mutual Understanding in a Global World.* (Cambridge: Cambridge University Press).

Breakwell, G. M. (ed.) (1992) *Social Psychology of Identity and the Self Concept.* (London: Surrey University Press).

Brunette, L., Bastin, G. L., Hemlin, I. and Clarke. H. (eds.) (2003) *The Critical Link 3: Interpreters in the Community.* Selected papers from the Third International Conference on Interpreting in Legal, Health and Social Service Settings. (Amsterdam/Philadelphia: John Benjamins Publishing Company).

Cambridge, J. (1999) 'Information loss in bilingual medical interviews through an untrained interpreter'. In I. Mason (ed.), *Dialogue Interpreting*, special issue of *The Translator*, 5 (2).

Campbell, S. and Hale, S. (2003) 'Translation and interpreting assessment in the context of educational measurement'. In G. Anderman and M. Rogers (eds.) *Translation Today: Trends and Perspectives* (Clevedon: Multilingual Matters).

Colin, J. and Morris, R. (1996) *Interpreters and the Legal Process*. (Winchester: Waterside Press).

Corbyn, J. (1989) 'Immigration Service (Interpreters)'. *Hansard, House of Commons*, 157, 6th series, col. 73.

Corsellis, A. (1992) 'Quality of service irrespective of language and culture'. *The Magistrate*, 92–3.

Corsellis, A. (1993a) 'How to work effectively with interpreters and translators'. *Criminal Practitioners Newsletter*, 14.

Corsellis, A. (1993b) 'A professional framework for linguists working in the UK legal system'. In C. Picken (ed.), *Translation – The Vital Link: Proceedings of the XIII FIT World Congress*, Vol. 1. (London: Chameleon Press).

Corsellis, A. (2008) *Interpreting for Public Services: The First Steps* (Basingstoke and New York: Palgrave Macmillan).

Crystal, D. (1997) *English as a Global Language*. (Cambridge: Cambridge University Press).

De Jongh, E. M. (1992) *An Introduction to Court Interpreting: Theory and Practice*. (Lanham, MD: University Press of America).

de Pedro Ricoy, R., Perez, I. and Wilson, C. (eds.) (2009) *Interpreting and Translating in Public Service Settings: Policy, Practice, Pedagogy*. (Manchester: St. Jerome Publishing).

Diriker, E. (2004) *De-/Re-contextualizing Conference Interpreting*. Amsterdam/ Philadelphia: John Benjamins).

Downing, B. T. and Helms Tillery, K. (1992) *Professional Training for Community Interpreters: A Report on Models of Interpreter Training and the Value of Training*. (Minneapolis: Center for Urban and Regional Affairs).

Drew, P. and Heritage, J. (eds.) (1992) *Talk at Work: Interaction in Institutional Settings*. Studies in Interaction Sociolinguistics 3. (Cambridge: Cambridge University Press).

Driesen, C. (1988) 'The interpreter's job – a blow-by-blow account'. In C. Picken (ed.), *ITI Conference 2 – Interpreters Mean Business*. (London: Aslib).

Driesen, C. (1989) 'Reformer l'interprétation judiciaire'. *Parallèles, Cahiers de l'École de Traduction et d'Interprétation de l'Université de Genève*, 11: 93–8.

Duranti, A. (1997) *Linguistic Anthropology*. (Cambridge: Cambridge University Press).

Edwards, A. B. (1995) *The Practice of Court Interpreting*. (Amsterdam/Philadelphia: John Benjamins Publishing Company).

Elder, C., Hargreaves, M., Kelmi, L. and Slatyer, H. (2006) 'An investigation into rater reliability, rater behaviour and comparability of test tasks'. Unpublished research report. (Sydney: Macquarie University and NAATI).

Erasmus, M. (ed.) (1999) *Liaison Interpreting in the Community*. (Hatfield, Pretoria: Van Schaik Publishers).

Ersozlu, E. 'Training of interpreters: some suggestions on sight translation teaching': http://www.translationdirectory.com/article755.htm; date accessed 12 May 2010.

Ervin-Tripp, S. M. (1982) 'Sociolinguistic rules of address'. In J. B. Pride and J. Holmes (eds.), *Sociolinguistics*. (Harmondsworth: Penguin).

Fairclough, N. (1995) *Critical Discourse Analysis: The Critical Study of Language*. (Harlow, Essex: Longman).

Fasold, R. (1990) *Sociolinguistics of Language*. (Oxford: Blackwell).

Finnegan, R. (2002) *Communicating: The Multiple Modes of Human Interconnection*. (London: Routledge).

Firth, A. (ed.) (1994) *The Discourse of Negotiation: Studies of Language in the Workplace*. (Oxford: Pergamon).

Foley, W. (1998) *Anthropological Linguistics: An Introduction*. (Oxford: Blackwell).

Fox, K. (2004) *Watching the English: The Hidden Rules of English Behaviour*. (London: Hodder).

Gaiba, F. (1998) *The Origins of Simultaneous Interpretation: The Nuremberg Trial*. (Ottawa: University of Ottawa Press).

Galanti, G.-A. (2002) *Caring for Patients from Different Cultures. Case Studies from American Hospitals*. (Philadelphia: University of Pennsylvania Press).

Garzone, G., Mead, P. and Viezzi, M. (eds.) (2002) *Perspectives on Interpreting. Papers from the First Forlì Conference on Interpreting Studies*. (Bologna: CLUEB).

Garzone, G. and Rudvin, M. (2003) *Domain-Specific English and Language Mediation in Professional and Institutional Settings*. (Milano: Arcipelago Edizioni).

Gentile, A., Ozolins, U. and Vasilakakos M. (1996) *Liaison Interpreting : A Handbook*. (Melbourne: Melbourne University Press).

Gibbons, J. (2003) *Forensic Linguistics: An Introduction to Language in the Justice System*. (Malden, MA: Blackwell Publishing).

Gile, D. (1995) *Basic Concepts and Models For Interpreter and Translator Training*. (Amsterdam: John Benjamins).

Gile, D. (ed.) (2001) *Getting Started in Interpreting Research*. (Amsterdam/ Philadelphia: John Benjamins Publishing Company).

Gillies, A. (2005) *Note-taking for Consecutive Interpreting: A Short Course*. (Manchester: St Jerome).

Goffman, E. (1981) *Forms of Talk*. (Oxford: Blackwell).

Gonzales, D. G., Vásquez , V. F. and Mikkelson, H. (1991) *Fundamentals of Court Interpretation: Theory, Policy, and Practice*. (Durham, NC: Carolina Academic Press).

Gumperz, J. (ed.) (1982) *Language and Social Identity*. (Cambridge: Cambridge University Press).

Hale, S. (1996) 'Bilingual encounters: Spanish-English medical and legal dialogues. A practical resource for educators and students of interpreting'. In Interpreting and Translation Publications Series, No. 1 (Western Sidney University, Macarthur. Language Acquisition Research Centre). Available at http://www.eric.ed.gov/PDFS/ED424759.pdf. Last accessed 20 July 2010.

Hale, S. (1999) 'The interpreter's treatment of discourse markers in courtroom questions'. *Forensic Linguistics. The International Journal of Speech, Language and the Law*. 6 (1): 57–82.

Hale, S. (2001) 'How are courtroom questions interpreted? An analysis of Spanish interpreters' practices'. In I. Mason (ed.), *Triadic Exchanges. Studies in Dialogue Interpreting*. (Manchester: St. Jerome).

Hale, S. (2004) *The Discourse of Court Interpreting: Discourse Practices of the Law, the Witness and the Interpreter*. (Amsterdam/Philadelphia: John Benjamins Publishing Company).

Hale, S. (2007) *Community Interpreting*. (Basingstoke and New York: Palgrave Macmillan)

Hale, S. (2008) 'Controversies over the role of the court interpreter'. In C. Valero-Garcés and A. Martin (eds.), *Crossing Borders in Community Interpreting*. (Amsterdam/Philadelphia: John Benjamins Publishing Company).

Hall, E. T. (1959/1984) *The Silent Language*. (New York: Doubleday).

Hall, E. T. (1976) *Beyond Culture*. (New York: Doubleday).

Hall, E. T. and Hall, M. (1983) *Understanding Cultural Differences. Germans, French and Americans*. (Boston: Intercultural Press).

Heimerl-Moggan, K. and John, V.-I. (2007) *Note-taking for Public Service Interpreters*. (Cheshire: Interp-Right Training Consultancy Ltd.)

Hertog, E. (2002) 'Language as a human right: the challenges for legal interpreting'. In G. Garzone and M. Viezzi (eds.), *Interpreting in the 21st Century: Challenges and Opportunities. Selected Papers from the First Forlì Conference on Interpreting Studies*. (Amsterdam/Philadelphia: John Benjamins Publishing Company).

Hertog, E. and van der Veer, B. (eds.) (2006) *Taking Stock : Research and Methodology in Community Interpreting*. (Antwerpen: Linguistica Antverpiensia series, Hoger Instituut voor Vertalers en Tolken).

Hertog, E. and van Gucht, J.(eds.) (2008) *Status Quaestionis Questionnaire on the Provision of Legal Interpreting and Translation in the EU*. (Mortsel: Intersentia Publishers).

Hofstede, G. (1991) *Cultures and Organizations. Software of the Mind*. (London: McGraw-Hill).

Hofstede, G. (2001) *Culture's Consequences: Comparing Values, Behaviors, Institutions, and Organizations Across Nations* (Thousand Oaks, CA: Sage Publications).

Hofstede, G. and Hofstede, G.-J. (2004) *Cultures and Organizations: Software of the Mind*. 2nd edition. (New York: McGraw-Hill).

Hofstede, G. and Pedersen, P. (2002) *Exploring Culture: Exercises, Stories*. (New York: Intercultural Press).

Holliday, A., Hyde, M. and Kullman, J. (2004) *Intercultural Communication: An Advanced Resource Book*. (London: Routledge).

Holmes, J. (2001) *An Introduction to Sociolinguistics*. (Harlow: Longman).

Hsieh, E. (2006) 'Understanding medical interpreters: re-conceptualizing bilingual health communication'. *Health Communication*, 20 (2): 177–86.

Hung, E. (ed.) (2002) *Teaching Translation and Interpreting 4*. (Amsterdam/ Philadelphia: John Benjamins Publishing Company).

Hymes, D. H. (1971) *On Communicative Competence*. (Philadelphia: University of Pennsylvania Press).

Jacobsen, B. (2002) 'Additions in court interpreting'. In G. Garzone and M. Viezzi (eds.), *Interpreting in the 21st Century: Challenges and Opportunities. Selected Papers from the 1st Forlì Conference on Interpreting Studies*. (Amsterdam/Philadelphia: John Benjamins Publishing Company).

Jaworski, A. and Coupland, N. (eds.) (1999) *The Discourse Reader*. (London: Routledge).

Jekat, S. (2002) 'On translation phenomena: reduction'. In G. Garzone and M. Viezzi (eds.), *Interpreting in the 21st Century: Challenges and Opportunities. Selected Papers from the 1st Forlì Conference on Interpreting Studies*. (Amsterdam/ Philadelphia: John Benjamins Publishing Company).

Jenkins, J. (2003) *World Englishes: A Resource Book for Students*. (London: Routledge).

Jones, R. (1998) *Conference Interpreting Explained*. (Manchester: St Jerome).

Kainz, C., Prunc, E. and Schögler, R. (eds.) (2010) *Modelling the Field of Community Interpreting: Questions of Methodology in Research and Training*. (London: LIT Verlag).

Kagitçibasi, Ç., Choi, S.-C. and Yoon, G. (eds.) (1994) *Individualism and Collectivism: Theory, Methods, and Applications*. (London: Sage Publications).

Katan, B. D. (2004) *Translating Cultures*. (Manchester: St. Jerome Publishing).

Kaufert, J. and Putsch, R. W. *et al.* (1997) 'Communication through interpreters in healthcare: ethical dilemmas arising from differences in class, culture, language and power'. *The Journal of Clinical Ethics*, 8 (1): 71–87.

Kleinman, A. (1980) *Patients and Healers in the Context of Culture. An Exploration of the Borderland between Anthropology, Medicine, and Psychiatry*. (Berkeley: University of California Press).

Kleinman, S. and Copp, M. A. (eds.) (1993) *Emotions and Fieldwork*. (London: Sage Publications).

Knapp-Potthoff, K., Knapp, A. and Enninger, W. (eds.) (1987) *Analysing Intercultural Communication*. (Berlin: Mouton).

Knuf, J. (1990) 'Greeting and leave-taking: a bibliography of resources for the study of ritualized communication'. *Research on Language & Social Interaction*, 24 (1): 405–48.

Krouglov, A. (1999) 'Police interpreting: politeness and sociocultural context'. In I. Mason (ed.), *Dialogue Interpreting*, special issue of *The Translator*, 5 (2).

Lakoff, R. T. (1989) 'The limits of politeness: therapeutic and courtroom discourse'. *Multilingua. Journal of Cross-Cultural and Interlanguage Communication*, 8 (2/3): 101–29.

Lundmark, T. (2009) *Tales of Hi and Bye. Greeting and Parting Rituals Around the World*. (Cambridge: Cambridge University Press).

Mason, I. (ed.) (2001) *Triadic Exchanges. Studies in Dialogue Interpreting* (Manchester: St Jerome Publishing).

Matsumoto, Y. (1989) 'Politeness and conversational analysis – observations from Japanese'. *Multilingua*, 8 (2/3): 207–21.

Merlini, R. (2009) 'Seeking asylum and seeking identity in a mediated encounter: the projection of selves through discursive practices'. *Interpreting*, 11 (1): 57–92.

Merlini, R. and Favaron, R. (2009) 'Quality in healthcare interpreting training: working with norms through recorded interaction'. In S. B. Hale, U. Ozolins and L. Stern (eds.), *The Critical Link 5*. (Amsterdam/Philadelphia: John Benjamins Publishing Company).

Metzger, M. (1999) *Sign Language Interpreting: Deconstructing the Myth of Neutrality*. (Washington: Gallaudet University Press).

Mikkelson, H. (2000) *An Introduction to Court Interpreting*. (Manchester: St. Jerome Publishing).

Moeketsi, R. (1999) *Discourse in a Multilingual and Multicultural Courtroom: A Court Interpreter's Guide*. (Pretoria: JL van Schaik Publishers).

Munday, J. (2008) *Introducing Translation Studies: Theories and Applications*. (London: Routledge).

Nida, E. (1964) *Toward a Science of Translating*. (Leiden: Brill).

Nida, E. and Taber, C. (1969) *The Theory and Practice of Translation*. (Leiden: Brill).

Niska, H. (1999) 'Testing community interpreters: a theory, a model and a plea for research'. In M. Erasmus (ed.), *Liaison Interpreting in the Community*. (Pretoria: Van Schaik).

Niska, H. (2002) 'Community interpreter training: past, present, future'. In G. Garzone and M. Viezzi, (eds.), *Interpreting in the 21st Century. Challenges and Opportunities*. (Amsterdam/Philadelphia: John Benjamins Publishing Company).

Niska, H. (2005) 'Training interpreters: programmes, curricula, practices'. In M. Tenent (ed.), *Training for the New Millennium*. (Amsterdam/Philadelphia: John Benjamins Publishing Company).

Niska, H. (2007) 'From helpers to professionals: training of community interpreters in Sweden'. In C. Wadensjö, B. Englund Dimitrova and A.-L. Nilsson (eds.), *The Critical Link 4*. (Amsterdam/Philadelphia: John Benjamins Publishing Company).

Numrich, C. (2002) *Hmong Shamanism and Hmong Health Care Choices*. Available at http://www.csh.umn.edu/Research/topics/hmongchc.html.

Phelan, M. (2001) *The Interpreter's Resource*. (Clevedon: Multilingual Matters).

Pöchhacker, F. (2004) *Introducing Interpreting Studies*. (London: Routledge).

Pöchhacker, F. and Schlesinger, M. (eds.) (2002) *The Interpreting Studies Reader*. (London and New York: Routledge).

Pöchhacker, F. and Shlesinger, M. (eds.) (2007) *Healthcare Interpreting: Discourse and Interaction*. (Amsterdam/Philadelphia: John Benjamins Publishing Company).

Pöchhacker, F., Lykke Jakobsen, A. and Mees, I. (eds.) (2007) *Interpreting Studies and Beyond. A Tribute to Miriam Shlesinger*. (Copenhagen: Samfundslitteratur).

Pollard, R. Q. Jr (1997–1998) *Mental Health Interpreting: A Mentored Curriculum*, University of Rochester. http://www.urmc.rochester.edu/deaf-wellness-center/products/mental-health-interpreting.cfm; date accessed 20 March 2010.

Psathas, G. (ed.) (1994) *Conversation Analysis: The Study of Talk-in-Interaction*. (Thousand Oaks, CA: Sage Publications).

Pym, A. (2010) *Exploring Translation Theories*. (London and New York: Routledge).

Rahman, T. (1999) *Language, Education and Culture*. (Karachi: Oxford University Press).

Renkama, J. (1993) *Discourse Studies. An Introductory Textbook*. (Amsterdam/Philadelphia: John Benjamins Publishing Company).

Riley, P. (2002) 'Epistemic communities: the social knowledge system, discourse and identity'. In G. Cortese and P. Riley (eds.), *Domain-specific English: Textual Practices Across Communities and Classrooms*. (Bern: Peter Lang).

Roberts, R. P., Carr, S. E., Abraham, D. and Dufour, A. (1998) *The Critical Link 2: Interpreters in the Community. Selected papers from the Second International Conference on Interpreting in Legal, Health and Social Service Settings*. (Amsterdam/Philadelphia: John Benjamins Publishing Company).

Roland, R. (1999) *Interpreters as Diplomats: A Diplomatic History of the Role of Interpreters in World Politics*. (Ottawa: University of Ottawa Press).

Rombouts, D. (2009) 'Interpreters and the police: do interpreters need to know interviewing techniques?' In *Aspects of Legal Interpreting and Translation*, Antwerp, Belgium. Available at: www.lessius.eu/tt/nieuws/eulita/default.aspx.

Roy, C. B. (1996) 'An interactional sociolinguistic analysis of turn-taking in an interpreted event'. *Interpreting*, 1 (1): 125–9.

Roy, C. B. (2000) *Interpreting as a Discourse Process*. (Oxford and New York: Oxford University Press).

Roy, C. B. (2006) *New Approaches to Interpreter Education*. (Washington, DC: Gallaudet University Press).

Roy, C. B. (ed.) (2009) *Innovative Practices for Teaching Sign Language Interpreters*. (Washington: Gallaudet University Press).

Rozan, J.-F. (1956) *La prise de notes en interprétation consécutive*. (Geneva: Georg). Trans. (2003) by A. Gillies as *Note-taking in Consecutive Interpreting*. (Cracow: Tertium).

Rudvin, M. (2002) 'How neutral is neutral? Issues in interaction and participation in community interpreting'. In G. Garzone, P. Mead and M. Viezzi (eds.), *Perspectives on Interpreting. Papers from the First Forlì Conference on Interpreting Studies*. (Bologna: CLUEB).

Rudvin, M. (2007) 'Taking stock of the situation: a critical review of community interpreting literature. Examining research paradigms and methodology'. In *Research and Methodology in Community Interpreting*, special edition of *Linguistica Antverpiensia*, New Series, 5: 21–41.

Russell, D. and Hale, S. (eds.) (2009) *Interpreting in Legal Settings* (Washington DC: Gallaudet University Press).

Russo, M. and Mack, G. (eds.) (2005) *Interpretazione di trattativa. La mediazione linguistico-culturale nel contesto formativo e professionale*. (Milano: Hoepli).

Salama-Carr, M. (ed.) (2007) *Translating and Interpreting Conflict*. (Amsterdam: Rodopi).

Sarangi, S.and Roberts, C. (eds.) (1999) *Talk, Work and Institutional Order. Discourse in Medical, Mediation and Management Settings*. (Berlin / New York: Mouton de Gruyter).

Sawyer, D. B. (2001) 'The integration of curriculum and assessment in interpreter education: a case study.' Unpublished doctoral dissertation, University of Mainz.

Sawyer, D. B. (2004) *Fundamental Aspects of Interpreter Education: Curriculum and Assessment*. (Amsterdam/Philadelphia: John Benjamins Publishing Company).

Schenkein, J. (ed.) (1978) *Studies in the Organization of Conversational Interaction*. (New York: Academic).

Scollon, R. and Wong Scollon, S. (1995) *Intercultural Communication: A Discourse Approach*. (Oxford: Blackwell).

Skaaden, H. (2007) 'Lexical knowledge and interpreter aptitude'. *International Journal of Applied Linguistics*, 9 (1): 77–97.

Skaaden, H. and Wattne, M. (2009) 'Teaching interpreting in cyberspace: the answers to all our prayers?' In http://www.stjerome.co.uk/books/c/2/ R. de Pedro Ricoy, I. Perez and C. Wilson (eds.) (2009), *Interpreting and Translating in Public Service Settings: Policy, Practice, Pedagogy* (Manchester: St. Jerome Publishing), pp. 74–88.

Slatyer, H. (2010) 'Integrating teaching, learning and assessment'. Paper presented at Critical Link 6, 26–30 July 2010, Aston University, Birmingham. (See http://www1.aston.ac.uk for abstract.)

Slatyer, H. and Carmichael, A. (2005) NAATI rater reliability report.Unpublished research report, Sydney: Macquarie University.

Spencer-Oatey, H. (2000) *Culturally Speaking. Managing Rapport through Talk Across Cultures*. (London: Continuum).

Spencer-Oatey, H. and Franklin, P. (2009) *Intercultural Interaction. A Multidisciplinary Approach to Intercultural Communication*. (Basingstoke and New York: Palgrave Macmillan).

Storti, C. (1994) *Cross-Cultural Dialogues: 74 Brief Encounters with Cultural Diversity.* (Yarmouth, NY: Intercultural Press, Inc.).

Storti, C. (2001) *The Art of Crossing Cultures.* (New York: Intercultural Press).

Strang, B. M. H. (1968) *Modern English Structure.* (London: Edward Arnold).

Tebble, H. (1999) 'The tenor of consultant physicians: implications for medical interpreting'. *The Translator,* 5: 179–200.

Tebble, H. (2003) 'Training doctors to work effectively with interpreters'. In L. Brunette, G. Bastin, I. Hemlin and H. Clarke (eds.), *The Critical Link 3: Interpreters in the Community.* (Amsterdam/Philadelphia: John Benjamins and Company).

Thiagarajan, S. (2004) *Simulation Games by Thiagi.* Workshop by Thiagi (Amherst: Hrd Press).

Thiagarajan, S. and Thiagarajan, R. (2006) *Barnga: A Simulation Game on Cultural Clashes.* (Boston: Intercultural Press).

Thierry, C. (1981) 'L'enseignement de la prise de notes en interprétation consécutive: un faux problème?' In J. Delisle (ed.) *L'enseignement de l'interprétation et de la traduction – de la theorie à la pédagogie.* Cahiers de traductologie 4. (Ottawa). pp. 99–112.

Toury, G. (1980) *In Search of a Theory of Translation.* (Tel Aviv: Porter Institute).

Toury, G. (1995) *Descriptive Translation Studies and Beyond.* (Amsterdam/ Philadelphia: John Benjamins and Company).

Tribe, R. (2003) *Working with Interpreters in Mental Health.* (New York: Brunner-Routledge).

Trompenaars, F. and Hampden-Turner, C. (1997) *Riding the Waves of Culture: Understanding Cultural Diversity in Business.* (London: Nicholas Brealey Publishing).

Trompenaars, F. and Hampden-Turner, C. (2000) *Building Cross-cultural Competence: How to Create Wealth from Conflicting Values.* (Chichester: John Wiley & Sons).

Valero-Garcés, C. (2007) 'Doctor-patient consultations in dyadic and triadic exchanges'. In F. Pöchhacker and M. Shlesinger (eds.), *Healthcare Interpreting.* (Amsterdam/Philadelphia: John Benjamins Publishing Company).

Valero-Garcés, C. and Martin, A. (eds.) (2008) *Crossing Borders in Community Interpreting: Definitions and Dilemmas.* (Amsterdam/Philadelphia: John Benjamins Publishing Company).

Vancouver Community College (2000) *Points of Departure. Ethical Challenges for Court and Community Interpreters.* (Video with teacher's manual). Vancouver: Open Learning Agency and Vancouver Community College.

Viaggio, S. (1995) 'The praise of sight translation (and squeezing the last drop thereout of)'. *Interpreters' Newsletter,* 6: 33–42.

Wadensjö, C. (1998) *Interpreting as Interaction.* (London and New York: Longman).

Wadensjö, C. (1999) 'Telephone interpreting and the synchronization of talk in social interaction'. *The Translator,* 5 (2): 247–64.

Wadensjö, C., Englund Dimitrova, B. and Nilsson, A.-L. (eds.) (2007) *The Critical Link 4: Professionalization of Interpreting in the Community: Selected Papers from the 4th International Conference on Interpreting in Legal, Health and Social Service Settings.* (Amsterdam/Philadelphia: John Benjamins Publishing Company).

Zhuang, E. (2009) 'Methods for developing intercultural conflict management skills'. Paper presented at the International Conference 'Costruire ponti verso il futuro', Modena, 8–9 October.

Websites (general)

Aegis Project (EU): http://www.aegis-project.eu; date accessed 20 July 2010.

http://www.businessballs.com/body-language.htm; date accessed 1 February 2010.

http://www.countrynavigator.com; date accessed 1 February 2010.

The Critical Link Conference (Conference on Community Interpreting held every three years): http://www.criticallink.org, date accessed 17 March 2010.

http://www.culturewise.net/training_coaching_solutions.html; date accessed 1 July 2010.

http://kkitao.e-learning-server.com/corpus/function.doc; date accessed 1 July 2010.

http://nccc.georgetown.edu/foundations/frameworks.html; date accessed 26 July 2010. This site hosts the video: *Infusing Cultural and Linguistic Competence into Health Promotion Training.*

Oslo University College: http://www.hio.no/content/view/full/4563 and http://www.imdi.no/en/Sprak/English/

http://www.web-us.com/memory/memory_and_related_learning_prin.htm; date accessed 11 July 2010.

Associations and training/accreditation programmes

CIES, interpreter training programmes for 'mediatori culturali'; information on immigration trends etc.:
http://www.cies.it/; date accessed 18 October 2009.

Interpreter Training Programme in Minnesota, information on community interpreting etc.:
http://www.cla.umn.edu/pti; date accessed 18 October 2009.

Human Rights Watch, useful information on refugee statistics, laws, news, etc.:
http://www.hrw.org/; date accessed 18 October 2009.

National Institute of Linguists PSI Diploma:
http://www.iol.org.uk/nrpsi/ default.asp; date accessed 18 October 2009.

NAATI Interpreter Accreditation Body, Australia:
http://www.naati.com.au; date accessed 3 June 2010.

Refugee Legal Center:
http://www.refugee-legal-centre. org.uk/; date accessed 3 June 2010.

Translatum, The Greek Translation Portal:
http://www.translatum.gr/journal/3/translator-glossary-en.htm

Unesco, Convention relating to the status of refugees:
http://www.unesco.org/most/rr4ref.htm; date accessed 3 June 2010.

UNHCR (United Nations High Commission for Refugees):
http://www.unhcr.ch/cgi-bin/texis/vtx/home; date accessed 3 June 2010.

Refugee statistics:
http://www.unhcr.ch/cgi-bin/texis/vtx/statistics; date accessed 3 June 2010.

Court interpreting

Aequitas Project (EU):
http://www.agisproject.com; date accessed 1 March 2010.

California Courts:
http://www.courtinfo.ca.gov; date accessed 1 February 2010.
http://courts.michigan.gov/lc-gallery/cornerstones.htm; date accessed 13 January 2010.
www.courttvcanada.ca; date accessed 13 January 2010.
Court TV News:
http://www.courttv.com/trials/famous/; date accessed 13 January 2010.
Law Dictionary:
http://dictionary.law.com; date accessed 13 January 2010.
Certification of interpreters and translators in Washington State:
http://www.dshs.wa.gov/trial/msa/ltc/itsvcs.html; date accessed 1 July 2010.
Immigration Appellate Authority (IAA), contains a page on the role of the interpreter:
http://www.iaa.gov.uk; date accessed 17 March 2010.
Immigration Advisory Service, contains 'Guidance Notes and Instructions for Casually-Employed Interpreters':
http://www.iasuk.org; date accessed 7 February 2010.
United Nations. International Criminal Tribunal for the former Yugoslavia:
http://www.icty.org.
Immigration Law Practitioners Association (ILPA):
http://www.ilpa.org.uk; date accessed 18 January 2010.
Institute for Translation and Interpreting (UK):
http://www.iti.org.uk; date accessed 13 March 2010.
EU interpreter standardization, Grotius project:
http://www.legalinttrans.info; date accessed 2 April 2010.
Forensic Linguistics. Now re-named *The International Journal of Speech, Language and the Law:*
http://www.equinoxpub.com/IJSLL
International Interpretation Resource Center (Monterey):
http://www.miis.edu/iirc/iirc2.html; date accessed 28 January 2010.
Australian Government. Migration Review Tribunal-Refugee Review Tribunal, Guidelines for Interpreters:
http://www.mrt-rrt.gov.au/Tribunal/InterpreterHandbook/17_Guidelines_for_interpreters.as; date accessed 28 January 2010.
Introduction to Court Interpreters:
http://www.courts.michigan.gov/scao/services/access/InterIntro.htm
Interpreter services and various information on interpreters; some demo videos
http://www.ninthcircuit.org/programs-services/court-interpreter/
Legal News:
http://news.findlaw.com; date accessed 1 February 2010.
National Register of Public Service Interpreters (UK):
http://www.nrpsi.co.uk; date accessed 14 June 2011.
United Nations. International Criminal Tribunal for the former Yugoslavia:
http://www.icty.org
Canada's National Crime Prevention Centre, NCPC:
www.publicsafety.gc.ca/prg/.../ncpc-about-eng.aspx; date accessed 4 April 2010.
Supreme Court of the United States:
http://www.supremecourtus.gov/oral_arguments/argument_transcripts.html; date accessed 25 March 2010.

Irish Translators' and Interpreters' Association (ITIA)
http://www.translatorsassociation.ie; date accessed 22 April 2010.
Manual of guidance for interpreters and police officers working with Interpreters:
http://www.west-midlands.police.uk/pdf/help.../Interpreter_Manual.pdf; date accessed 20 July 2010.

Interpreter/translation associations

Association International des Interprètes de Conférence:
http://www.aiic.net; date accessed 2 April 2010.
http://www.aiic.net/ViewPage.cfm/page54.htm (this page contains the AIIC code of ethics); date accessed 1 April 2010.
International Association of Professional Translators and Interpreters:
http://www.aipti.org/eng/; date accessed 11 July 2010.
Association of Police and Court Interpreters:
http://www.apciinterpreters.org.uk; date accessed 14 March 2010.
American Translators Association (information on accreditation, and more):
http://www.atanet.org; date accessed 2 April 2010.
Australian Institute of Interpreters and Translators Inc.:
http://www.ausit.org; date accessed 18 October 2010.
This page contains the code of ethics:
http://www.ausit.org/ethics.php; date accessed 18 October 2009.
This page contains Assessment Criteria for an Interpreting Job:
http: www.ausit.org/pics/Asscrittranslation05.doc; date accessed 18 May 2010.
Vancouver Community College (training programmes for public service interpreters):
http://continuinged.vcc.ca/interpreting/index.htm; date accessed 18 June 2010.
American Sign Language Providers:
http://fortress.wa.gov/dshs/maa/InterpreterServices/ASLproviders.htm; date accessed 18 October 2009.
International Medical Interpreters Association (IMIA):
http://www.imiaweb.org; date accessed 20 July 2010.
National Association of Judiciary Interpreters and Translators (US):
http://www.najit.org/Publications/Proteus_index.htm; date accessed 22 January 2010.

Medical interpreting

Macmillan Cancer Relief:
http://www.cancerlink.org; date accessed 30 January 2010.
Resources for Cross Cultural Health Care (US)
http://www.diversityrx.org; date accessed 15 May 2010.
Australian Government Department of Health and Ageing:
http://www.health.gov.au; date accessed 13 June 2010.
IMIA medical training conference, July 2010:
http://www.imiaweb.org/conferences/2010conference.asp; date accessed 1 July 2010.
Australian Department of Immigration and Multicultural Affairs:
http://www.immi.gov.au; date accessed 1 April 2010.

California Healthcare Interpreters Association:
http://www.interpreterschia.org; date accessed 12 March 2009.
Language Line Ltd. Telephone Interpreting Service in the UK and US:
http://www.languageline.co.uk ; date accessed 20 January 2010.
Medics on the Move:
http:// www.medicsmove.eu; date accessed 1 July 2010.
NSW Multicultural Health Communication Service:
http://www.mhcs.health.nsw.gov.au; date accessed 1 April 2010.
International Medical Interpreters Association:
http://www.mmia.org/; date accessed 20 January 2010.
National Council on Interpreting in Health Care (USA):
http://www.ncihc.org; date accessed 25 April 2010.
Reach Nola (New Orleans)
http://reachnola.org/pdfs/Sight%20Translation%20and%20Written%20Translation.
    pdf; date accessed 29 May 2010.
Scottish Interpreters and Translators Association (SITA).
http://www.si-ta.org; date accessed 20 July 2010.
National Union of Professional Interpreters and Translators (NUPIT) (UK)
http://www.unitetheunion.org/nupit; date accessed 20 April 2010.
Cross-cultural Health Care Programme in Seattle, the US:
http://www.xculture.org; date accessed 2 April 2010.
Medical expressions English:
http://www.dh.gov.uk/prod_consum_dh/groups/dh_digitalassets/@dh/@en/
    documents/digitalasset/dh_4073282.pdf
Medical expressions in Italian:
http://www.dh.gov.uk/prod_consum_dh/groups/dh_digitalassets/@dh/@en/
    documents/digitalasset/dh_4073451.pdf

Telephone interpreting

http://www.languageline.com/www.languageline.uk/; date accessed 1 June 2010.
http://www.networkomni.com; date accessed 1 June 2010.

Videoconferencing

AT&T language services for video conferencing:
http://www.att.com/conferencing/vid_bas.html; date accessed 1 June 2010.

Books and books with CD-roms used in class

Eric H. Glendinning and Beverly A.S. Holmström, *English in Medicine. A Course
    in Communication Skills* (2004) Audio CD, (2005*) Coursebook*, 3rd edition.
    (Cambridge:Cambridge University Press).
Barbara Bettinelli, Paola Catenaccio and Karine Beatty (2006) *English for Medicine*,
    volumes 1 and 2. (Milano: Ulrico Hoepli Editore).
Oxford English for Careers (various publication dates) *Commerce 1, Commerce
    2, Medicine 1, Nursing 1, Nursing 2, Technology 1, Tourism 1, Tourism 2, Tourism 3*
    Oxford: Oxford University Press.

Brochures/information documents used in class

American Environmental Health & Safety:
http://www.healthsafety.com
British Heart Foundation:
http://www.bhf.org.uk/
Child Abuse Prevention Council Coalition (USA):
http://www.sierrasaccoalition.org/
National Heart, Lung, and Blood Institute (USA):
http://www.nhlbi.nih.gov/
The National Osteoporosis Society (UK):
http:// www.nos.org.uk/
The Stroke Association (UK):
http://www.stroke.org.uk/
UK Department of Health:
http:// www.dh.gov.uk/
US Department of Health and Human Services:
http://www.hhs.gov/
Useful sites for Medical English:
http://www.englishmed.com/ (Malta)
http://www.freewebs.com/medlinemalatya/index.htm

DVDs and videos used in class

Unethical? Who me?
AUSIT, 23 September 2004, Casa d'Italia (Co.As.It.) Leichhardt, NSW, Australia.
Dealing with Stress while Interpreting
AUSIT, 20 April, 2006, Casa d'Italia (Co.As.It.) Leichhardt, NSW, Australia.
Interpreting and Translating for the Police (2)
AUSIT, 8 February, 2007, Casa d'Italia (Co.As.It.) Leichhardt, NSW, Australia.
http://courts.michigan.gov/scao/services/access/InterIntro.htm
Arraignment (4:45 min)
Bail Hearing (1:10 min)
Sentencing (2:53 min)
Trial (7:40 min)
Points of Departure: see Vancouver Community College (2000).
Mental Health Interpreting: A Mentored Curriculum: see Pollard (1997–8).

Index